Managing the Mean Math Blues

Author and Illustrator
CHERYL OOTEN

SANTA ANA COLLEGE

Illustrator
EMILY MEEK

...er, New Jersey
Columbus, Ohio

Library of Congress Cataloging-in-Publication Data

Ooten, Cheryl.
    Managing the mean math blues / Cheryl Ooten ; illustrator Emily Meek.
      p. cm.
    Includes bibliographical references and index.
    ISBN 0-13-043169-9
    1. Mathematics—Study and teaching—Psychological aspects.   I. Title.

QA11.2 .O67 2003
510'.71—dc21   2002025139

Vice President and Publisher: Jeffery W. Johnston
Senior Acquisitions Editor: Sande Johnson
Assistant Editor: Cecilia Johnson
Production Editor: Holcomb Hathaway
Design Coordinator: Diane C. Lorenzo
Cover Designer: Jeff Vanik
Cover Art: SuperStock
Production Manager: Pamela D. Bennett
Director of Marketing: Ann Castel Davis
Director of Advertising: Kevin Flanagan
Marketing Manager: Christina Quadhamer

> *To my husband, Bob,*
> *for his support and patience;*
> *to my first algebra teacher,*
> *my father, R. J. Thomas;*
> *and to my students*
> *for being brave*
> *and working hard.*

This book was set in Janson by Aerocraft Charter Art Service. It was printed and bound by R. R. Donnelley & Sons Company. The cover was printed by Phoenix Color Corp.

Pearson Education Ltd.
Pearson Education Australia Pty. Limited
Pearson Education Singapore Pte. Ltd.
Pearson Education North Asia Ltd.
Pearson Education Canada, Ltd.
Pearson Educación de Mexico, S.A. de C.V.
Pearson Education–Japan
Pearson Education Malaysia Pte. Ltd.
Pearson Education, *Upper Saddle River, New Jersey*

10 9 8

ISBN 0-13-043169-9

Brief Contents

Part Four Do the Math   99

Contents

v

Part Two   Change Your Thoughts   27

Part Three   Know Yourself   63

Preface

Dear Reader,

My fondest wish is that this book will assist you to succeed with math. Feel free to read it in any order that works for you. This is a book for you to control. The techniques and exercises are here to help you, not to overwhelm or discourage you. If you feel overwhelmed or discouraged, back off and return later. But do return. The rewards are many and great.

I have included information I have found useful to math students during 30 years of teaching, so pick and choose. Refer to this book when you need a new and different strategy.

Dawn Bigelow, a superb third-grade teacher I taught beside, told me that when she went to a conference she wanted to return with *three new ideas*. More than three and she would be too overwhelmed to try them. Fewer than three and she had wasted her time going to the conference.

Three was the magic number. When Dawn returned to her third-graders with three new ideas, she could easily incorporate them into the classroom system she already had in progress.

You have a system in progress for learning. You only need three new ideas each time you come to this book. More than that and you will be overwhelmed. Fewer than that and you will be wasting your time. Modify your learning system slowly and surely. Incorporate winning ideas and strategies that fit who you are and what you want to accomplish.

Skim over the Contents. Mark the topics that look the most promising. Chapter 2, along with the list below, can direct you according to your needs.

Features of this book and their purpose are:

- Introduction and Chapter 1: Motivation to excite you and help you gather courage to confront the blues.
- Chapter 2: Explanation of routes through this book based on your needs.
- Chapters 3–6: Effective methods to control overwhelming negative thoughts and feelings about math (or life).
- Chapters 7–9: Self-discovery about who you are and how you learn best.
- Chapters 10, 13, 14, and 15: Study skills to use in math class.
- Chapters 11 and 12: Discussion of shyness and classroom/teacher issues.
- Chapter 16: Problem-solving strategies.
- Chapter 17: Test-taking strategies.

- "Pushing Your Limits," Chapters 1–17: Journal activities to help you question, ponder, plan, and evaluate your math life.
- "Mastering Math's Mysteries," Chapters 3–6: Practice with numbers and patterns.
- "Mastering Math's Mysteries," Chapters 7–13: Fraction practice to shore up skills that math students tend to avoid.
- "Mastering Math's Mysteries," Chapter 15: Practice with spatial visualization.
- "Mastering Math's Mysteries," Chapters 14, 16, and 17: Practice with strategies discussed in the chapters.
- "More Mastering Math's Mysteries," in the Appendix: More challenging math practice for the brave of heart.
- Solutions to "Mastering Math's Mysteries" exercises, in the Appendix.

This book is not designed as a math textbook but rather to accompany a math textbook or to prepare you for a math textbook. The math exercises here are just for you to wet your feet. Because I know that every math student brings different experiences and needs, I had difficulty deciding which math topics to include. I chose fractions because they are universally avoided and disliked. The potential exists for you to feel terrific soon if you face them. Be patient with yourself as you wade into new territory. Being curious and willing to experiment can help you to swim sooner than you ever thought possible. Use a life preserver when you need it and never swim alone.

My best,

Cheryl Ooten

Acknowledgments

I wish to thank the following people:

- Bobbi Nesheim; Melba Finklestein; my daughter, Kati Ooten Parker; Jack Porter; Ruth Afflack; and my parents-in-law, Blanche and LaVerne Ooten, who inspired me and gave me courage.

- Virginia Starrett; my brother-in-law, Bill Browne; my sister, Linda Browne; Emily Meek; Kathy Moore; Carleen Ono and her Writing Group; Maureen Pelling and her Writing Group; my mom, Doris Thomas; and my editor, Sande Johnson, who read my work and spurred me on.

- The reviewers who also read my work an offered constructive suggestions and encouragement: Monica Geist, Front Range Community College; Hugh Horan, New Mexico Highlands University; Rhonda MacLeod, Tallahassee Community College; Marilee McGowan, Oakton Community College; Valerie H. Maley, Cape Fear Community College; Kathleen Miranda, SUNY College at Old Westbury; Elsie Newman, Owens College; Faustine Perham, University of Wisconsin at Whitewater; Fred Peskoff, Borough of Manhattan Community College; Christopher Reisch, Jamestown Community College; Jennifer Sawyer, Currituck County Schools; Paul E. Seeburger, Monroe Community College; Lymeda Singleton, Abilene Christian University; and Joan Totten, Ferris State University.

- Mary Anne Anthony, Lynn Marecek, Kathy Taylor, Christa Machir, the rest of my dynamic department, Mary Halvorson, Melody Vaught, Dennis Gilmour, Mike Petyo, Tim Cooley, Russ Meek, Karin Wright, Victoria Stephenson, James Palmer, Bryan Kehlenbach, David Wintle, and other friends and colleagues who shared their ideas, expertise, and experience.

- Bob White, Don McIntyre, Giovanna Piazza, Emily Meek, Bobbi Nesheim, Elinor Peace Bailey, Judy Schaftenaar, Phyllis Biel, Joel Sheldon, Jazmin Hurtado, Isabella Vescey, Enrique De Leon, Sarah Kershaw, and Alex Solano, who granted me interviews and spoke with candor.

- My drawing teachers, Emily Meek and Bob White.

- Santa Ana College, for granting me a sabbatical to read, interview, and write.

Limits
Limits
Limits
Limits
Limits

Introduction

DIANE ARBUS   "My favorite thing is to go where I've never been."

You are, no doubt, reading this book because of troubles you face with your mathematics studies. These difficulties may seem beyond your control and imposed upon you.

I would like you to consider that these difficulties may be within your circle of influence. Working through this book may give you the

- information
- courage
- support

to push your math skills further than you now believe possible.

We all have limits. I believe that we just don't know what they are. Often we think our limitations are larger than they truly are. I believe that about you and math. Choosing to take charge and accept responsibility will allow you to push past the limitations that you have set for yourself.

THE COMPANY YOU KEEP

- Award-winning **woodworker Sam Maloof** pushed the limits. At the age of 32 he decided to become a woodworker. Except for working as a graphic artist, Maloof was not formally trained in art and did not know any other woodworkers. He pushed the limits for furniture making. Still active in his

If you choose to push your limits, you will be in good company.

eighties, examples of his creations reside in numerous art museums as well as the White House. Maloof (1983) describes his experience:

> With the upholstery, I would figure out what I wanted to do and tell the upholsterer. **He would say, "It can't be done." And I would answer, "Why don't we try it anyway?"** Then very reluctantly he would do what I asked him. For a long time, however, when he would see me drive up to his shop, he would disappear. (p. 186)

- **Physicist Lise Meitner** (1878–1968) was told in the early 1900s that women were not allowed in the chemistry labs nor could they attend chemistry lectures. **She did not let that stop her.** She made her own lab in the basement of the chemistry building and hid behind the furniture in the back of the lecture hall during presentations. Today she is known as one of the greatest minds behind discoveries in radioactivity and nuclear physics.

- Sixty-year-old **artist and doll creator Elinor Peace Bailey** pushes the limits as she travels the world teaching women how to create dolls that reflect real people and real emotions. Dressed like one of her doll creations, with short purple hair, she starts her presentations on drawing faces with the statement, "In one day I cannot teach you what I have learned artistically since I drew on my parents' walls as a child. **What I want to do is to change your mind. It is not 'I can or cannot draw.' It is 'I *do* or *do not* draw.'"**

• The brilliant **physicist Albert Einstein** (1879–1955), labeled slow and disruptive as a youth in school, said, **"Imagination is more important than knowledge."** As an adult, Einstein used his imagination to revolutionize physics.

• **Mathematician Emmy Noether** (1882–1935), because of her gender, received no pay and lectured under the name of famous mathematician David Hilbert at the University of Gottingen. Recognized as changing algebra methods forever, **Noether persisted with creating and discussing new algebra** with students and colleagues, leaving an extraordinary mathematical legacy.

These five people looked at their limits and the belief systems around them. Then they pushed further. You will meet more people in this book who have pushed their limits. They are people like you who seek a goal beyond their previous expectations.

To push your limits, acknowledge the limitations in your life— then challenge them.

Reading this book will help you discover limits you have placed on yourself in math. As you recognize them, you can challenge and conquer them.

Lay the Groundwork

The Mean
Math Blues

ANAIS NIN   "Life shrinks or expands according to one's courage."

As a math teacher, countless students tell me:

- "In fourth grade, my teacher began fractions. I was clueless. Every time I see a fraction now, my mind closes."
- "I loved math in elementary school. As I started algebra, everything changed. Math made no sense anymore."
- "I do and understand my homework. In math exams, I forget everything."
- "In class, I follow the teacher. At homework time, the problems are beyond me."
- "In high school, my math teacher seemed more interested in coaching. He ran our class competitively and I didn't measure up. I gave in and failed."
- "I avoid numbers. After lunch with friends, I hope someone else will figure what I owe."
- "All my life I've dreamed of being a teacher. The problem is that Liberal Arts Math is required and I can't pass the prerequisites."
- "I know there are math problems on my state licensing exam. I am terrified I will fail."

These people experience what I call the "Mean Math Blues" or what some people call "Math Anxiety."

Signs of the Mean Math Blues

anger	fear	helplessness	avoidance
sadness	confusion	boredom	frustration
nausea	shaking	depression	palpitations
hives	tension	headaches	shame

Because you are reading this book, you have likely thought some of the same things my students tell me. You might possibly have experienced a few of the signs listed or others.

YOU ARE NOT ALONE!

More than one third of my math students share similar feelings and experiences. In survey after survey from basic math through calculus, students tell me they experience anxiety about math or tests. As a math teacher, this concerns and saddens me. I know my math colleagues feel the same. We all love math and find it, for the most part, exciting and challenging. We also want to share that experience of finding joy in math with our students.

I Got the Blues

What might make me different from many of my math colleagues is that, in the fall of 1967, during my second year of math graduate school, I experienced what I can now describe as a full-blown case of math anxiety or the Mean Math Blues.

During high school math, I was a "big fish in a small pond." I could figure out the algebra, geometry, and trigonometry problems on my own. Since I was one of few in my graduating class of 60 students who did the homework, I seldom had much competition, which gave me an inflated view of my math skills. However, I was enthralled by the patterns, interrelationships, and progressions of the math work. When fellow student Linda Nielsen said that I explained geometry better than the teacher did, I was hooked and looked forward to explaining math to students as a profession.

Even though my undergraduate math courses in college forced me to pore over the book and spend countless hours at my desk deciphering my notes to work homework, my previous success in high school and the certainty of teaching math in the future cheered and strengthened my resolve. I believed that no matter how difficult the material became, I could eventually understand.

A year after receiving a bachelor's degree in math, I returned to school to work on a master's degree in the same. My second year, I took a course called "Uniform Spaces" and, frankly, I never discovered exactly what a "uniform space" was.

Each Tuesday and Thursday morning of that fall, I woke feeling dread as I realized I had to go to "Uniform Spaces" class. This was a new and debilitating experience for me. I sat in front desperately trying to write everything said or written on the board. I felt my brain turn to cotton. The professor's voice came from far away. I thought myself too shy to ask fellow students for assistance with questions and homework. There was no textbook, so my notes were all I had to study.

As the semester dragged on, I felt more and more relief as class was dismissed. I avoided homework until, you guessed it, right before the next class when my high anxiety level prevented clear thinking. The exams were take-home exams. I spent several unpleasant weekends alone combing my notes and brain to do the problems.

Somehow I managed to pull a B grade. I am not sure how. I believed a B grade was adequate and was happy with B's before. When I chose to be a math major, I knew I would not earn all A's. But this was different. I'd lost confidence and endured a thoroughly miserable semester with what I thought was my true academic love and professional calling. It took me years of feeling blue to understand how I could have changed my behavior and the whole experience for the better.

How Do You Spell Relief?

In 1978, the title of Sheila Tobias's book *Overcoming Math Anxiety* on the shelf in the Glendale Galleria bookstore jumped out and filled me with relief. I finally had words of explanation for my "Uniform Spaces" ordeal.

Those two little words—math anxiety—helped me separate myself from my experience, give it a name, and take control over it. I read everything I could find about anxiety and math—discovering many ways I could have thought and behaved differently 10 years earlier. Since 1980, one of my great pleasures has been sharing my discoveries in classes and workshops with students who are math anxious.

Mean Math Blues Boogie

In the fall of 1998, while I was driving home from a refreshing weekend in the mountains, I was listening to my favorite jazz radio station, KLON. A blues song came on, catching my attention. I started playing with the words and melody and came up with the phrase "Mean Math Blues" and a 16-bar blues boogie. The words "Mean Math Blues" seemed to cover just about all of the difficulties math students encounter. Since the blues come and go, "managing" them promised hope and control without raising the expectation of a total cure. Singing about them invoked a playful spirit that I hoped could infuse my students. The song is in the Appendix.

YOU CAN MAKE NEW CONNECTIONS

Recent scientific discoveries about the human brain verify the limitless potential for new learning and new brain connections at all ages. These discoveries promise new possibilities for those who are willing to "draw outside the lines."

There are many ways to do "something different" as you work with math. Regardless of the causes or the form of your Mean Math Blues, this book can help you:

- Know more about yourself.
- Learn new math strategies.
- Mobilize your support systems and resources.
- Take charge of your thoughts and behaviors.

It is possible to change avoidance in math to excitement about conquering challenges. Conquering the Mean Math Blues is a process—a process that takes intention and energy. *It will not happen overnight.* Sometimes you will progress quickly. Sometimes, slowly. You need to face only as much as you can handle at one time. As you work through your current issues, you will have more energy to move on to new ones.

Because you are reading this book, I know that

- You, like my students, also push barriers.
- You examine limiting expectations.
- You challenge yourself and won't settle for less than your best.
- You actively educate yourself in order to gain opportunities for a limitless life.

I invite you to continue this journey as you read and as you conquer the math challenges of your choice. Examine those limiting beliefs and consider the outcomes, then jump in—take responsibility and choose an expansive way of living. You will not regret it and you can expect your life to be a series of amazing events.

2

A Place to Begin

ARCHIMEDES   "Give me where to stand, and I will move the earth."

9

The Climb

Imagine watching a rock climber. Notice how carefully she examines the route ahead to make choices for the next move. She has four contacts with the rock—with her two hands and her two feet.

A basic rule for rock climbing is to move only one contact at a time. With one hand or one foot, she experiments with the next move that will take her up and keep her secure. **The important issues for her are controlling the situation and continuing to move along a route with positive options.**

Be like the rock climber. Take control, keep your solid contacts with the ground, and move in directions with positive options. Every once in a while, glory in looking back to see how far you have come.

Then this will be a safe journey for you because you will choose the rocks that you are willing to climb and you will take charge of each moment. Your journey will progress one move at a time. Your focus will be on the "here and now" as you choose from your currently available options.

HOW TO USE THIS BOOK

Every reader has different wants and needs. These chapters do not have to be read in order. Some readers will be in a math class and need specific assistance with current issues. Other readers will be thinking about taking a math course and need their courage bolstered. All readers will find far more in this book than they can possibly take in and use. A few good ideas make the difference between success and failure.

Keep this book as a reference tool that can serve your changing needs as you study and use math. Take advantage of the strategies. Actively read and do the exercises. Recognize the many ways you can control your math experience. Experiment with different ideas to find what works best for you.

ROUTES TO SUCCESS

Path #1

If you are in a math class now and need specific suggestions immediately, read Chapters 10–17 first. Then come back to Chapters 3–9 to know more about yourself. Place a check mark beside topics that are urgent to you in the following list.

- For overall success tips, read the Epilogue.
- For test preparation and test taking, read Chapter 17.
- For coping with test results, read "Refocus after the Exam" in Chapter 17.
- For remembering math more accurately and longer, read Chapter 13.
- For feeling more successful as you work with math, read Chapter 10.
- For day-to-day study strategies, read Chapters 10 and 15.
- For words from other math students, read Chapters 9 and 15.
- For perspective about classrooms and teachers, read Chapter 12.
- For successful problem-solving strategies, work through Chapter 16.
- For a dynamic, visual method of organizing or brainstorming ideas, see Chapter 14.

Path #2

If you have overwhelming negative thoughts or fears about math, continue reading here and read Chapters 3–6 first. Then go on to the rest of the book. Experiment with the Mastering Math Mysteries exercises *slowly* to build your confidence, since avoidance of math only makes the negative emotions bigger. Take your time and, paradoxically, you will progress more quickly.

- For immediately checking out your math beliefs, read Chapter 3.
- For help with feelings and physical symptoms, work Chapters 4, 5, and 6.
- For changing negative thinking, work Chapters 5 and 6.
- To know more about yourself as a learner and how to maximize your strengths in math, read Chapters 7 and 8.
- For encouraging words and experiences of others, read Chapters 9 and 15.

TAKING CHARGE

*Anxiety results when you are **required** to stay in an uncomfortable situation where you **believe** you have no **control.***

As a child, you depended upon the adults in your life and were *required* to participate in many experiences—in particular, math experiences—whether you liked it or not. Fortunately, as an adult, you have more choices. Although you cannot control everything happening in your life, if you choose to do so, there is much you can control.

As you choose a major or vocation with a math component, you place yourself voluntarily in a math class. You are now choosing the math requirement. Every single time you attend class, remind yourself that you have chosen this hard work to gain something you want—the profession of your choice.

Your beliefs affect your actions. If you *believe* you cannot do math, you won't experiment with it or practice it or place yourself where you can learn more. If you *believe* people who are "good" in math do math quickly, you will be impatient with yourself when you learn. You won't allow yourself the necessary "percolating" time for math to settle into your mind. (Chapter 3 will challenge more of your beliefs.)

You assume *control* now by reading this book, by answering the questions, by examining your previous math habits, by being honest with yourself, by experimenting with math problems at your level, by recognizing your right to base your self-esteem on things other than your math skills, by asking questions when you don't understand . . .

*A student once told me that her breakthrough in math came when she realized she wanted to be a teacher **more** than she was afraid of math.*

As you examine your math beliefs in this book and change requirements to choices, you take charge of your math life and your anxiety level will decrease.

CHARTING YOUR PROGRESS

Begin a journal in which you write your experiences and thoughts as you read this book and as you work with math.

Purchase a notebook or sketchpad to use. Write down your thoughts, ideas, goals, feelings, doodles, notes, lists—anything that comes to your mind during the day that might involve you and math *or* you and learning *or* you and ?. Remember: **Change is a process. Change takes time.** Date your entries so you can see the progression. Your journal—a powerful tool for change—will help you become more conscious of your inner thoughts. **Play with your journal and see what happens.**

I use an artist's sketchpad to doodle, take notes, and copy favorite quotations, as well as to write my personal journal. The writing does not have to be perfect. In fact, the more you allow the words to flow uncensored from your pen, the more you will learn about yourself and be able to release beliefs that do not work for you anymore.

Sometimes as I write in my journal, I think that what I am writing is trivial and a waste of time. I choose to ignore that thought and keep writing. Later I often discover that the writing process has revealed something new *or* I discover that I feel differently after writing and I am released to go on to important work that I want to do. My journaling provides me with direction and focus.

I experimented with different colored pens and pencils and discovered that changing colors and writing tools helped me move beyond my previous limitations. In fact, this experimentation helped me discover some unrecognized talent for drawing despite my image of myself as a non-artist. The drawings beginning the chapters in this book are mine.

PREVIEW: PUSHING YOUR LIMITS

At the end of each chapter is a section called "Pushing Your Limits." These exercises can help you know more about yourself and your abilities so that you can push your barriers further. **Enjoy the exercises. Work them at your own pace and at your own level. Read all of the exercises and choose to do those that fit your current situation. Return and complete the other exercises later when you are ready for more growth.**

Pushing Your Limits

1. Write down your math wish list in your journal. What are your goals for yourself? What do you want to accomplish in math? Do you want more life skills such as handling money more easily or understanding the percentages in the daily newspaper? Do you want to complete certain math courses? Do you have math ideas that you would like to clarify?

2. Think about the sentence, "Anxiety results when you're required to stay in an uncomfortable situation where you believe you have no control." How can you take more control now? (Exercises 3, 4, and 5 suggest possibilities.) Leave space in your journal so that you can return to this exercise. You will have more to add as you work through this process. Remember to date your journal entries.

3. Begin to experiment with changes in your vocabulary. When you say, "I have to . . ." or "I should . . ." or "I must . . .," change to "I choose to . . ." When you are tempted to say, "They are making me . . .," change to "I choose to . . ." Write about this process.

4. Choose one of these goals to work on this week in your math class and evaluate your progress one week from today:

   a. Work three review problems per day to boost your confidence.

   b. Cheer yourself as you learn new ideas. Smile in class.

   c. Mark where you don't understand your notes and textbook.

    Make a "Goals Page" in your journal to record this goal and more later. Document your progress each week.

5. Choose one of these actions to get more feedback on your understanding of math this week. Make an "Actions for Feedback" page in your journal.

   a. Summarize the procedures and main ideas of a problem.

   b. Teach someone else how to do homework problems.

   c. Ask questions in or out of class.

At the end of the week, decide if this helped you understand your math problems more easily. Record your decision.

3

Challenge Your Beliefs

"It is well to remind ourselves that anxiety signifies a conflict, and so long as a conflict is going on, a constructive solution is possible."

ROLLO MAY

Are You Cutting Yourself Short?

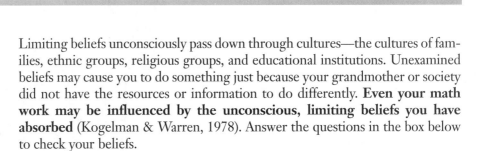

A young woman questioned her mother, who was preparing the traditional roast for the family holiday celebration. Every year the mother cut off both ends of the roast and put all the pieces into a pan in the oven. "Why do you cut off the ends?" the young woman asked. The mother responded that *her* mother had always done that. When the young woman questioned her grandmother, the grandmother said she cut off the ends because she had never owned a pan large enough to hold an entire roast.

Limiting beliefs unconsciously pass down through cultures—the cultures of families, ethnic groups, religious groups, and educational institutions. Unexamined beliefs may cause you to do something just because your grandmother or society did not have the resources or information to do differently. **Even your math work may be influenced by the unconscious, limiting beliefs you have absorbed** (Kogelman & Warren, 1978). Answer the questions in the box below to check your beliefs.

Quiz

Answer True or False before reading on to learn about your math beliefs.

_____ 1. I can't do math.

_____ 2. Math is always hard.

_____ 3. Only smart people can do math.

_____ 4. Mathematicians always do math problems quickly in their heads.

_____ 5. If I don't understand a problem immediately, I never will.

_____ 6. There is only one right way to work a math problem.

_____ 7. I am too shy to ask questions.

_____ 8. It is bad to count on my fingers.

_____ 9. To complete my math requirements quickly, I should skip to the highest level math class that I can.

_____10. My memories of my negative math experiences will never go away.

Each statement on the previous page is *false*. **Are you surprised?** These myths are negative math beliefs acquired subtly through cultures, families, teachers, and friends. For math students, negative beliefs can be deadly. At the very core of your self-esteem, these negative beliefs keep you uneasy.

Belief in these myths is a stumbling block to learning math because you think and act from your belief system.

CHALLENGE YOUR NEGATIVE BELIEFS

I Can't Do Math

Not true! People who can count, add a few numbers, recognize circular and rectangular shapes, and point to the front and back of the classroom can do math. All these skills are math.

You do math all the time. Knowing your age, comparing sizes and shapes, adding your money, and subtracting to get change are math skills. You use math every day of your life at home and at work without giving it a second thought.

You drive the streets judging distances, speeds, and times. You estimate if you can afford a vacation or a car and when you can retire. You compare volumes as you cook and areas as you rearrange furniture. You measure volume as you put toothpaste on your toothbrush. You use statistics as you watch sports and consider things like RBIs in baseball or field goal percentages in basketball. All of these are mathematical skills taken for granted.

Often students enroll in math classes beyond their skill level and become discouraged. **The way to regain your confidence is to slow down and discover the level of math where you learn new concepts, but are not overwhelmed.**

Math Is Always Hard

There are some math concepts that are hard for everyone. There are math problems that have been unsolved for hundreds of years even though they've been attempted by competent, knowledgeable mathematicians who may work at them for decades. Those aren't the problems you need to work, unless you are curious. When you work at your appropriate level, you find a combination of easy ideas and hard ideas.

You may get discouraged when you compare your speed and understanding in math with that of your teachers. Math teachers appear to do math easily because

they have done these specific problems, or ones like them, many, many times. Also the problems you see them work in class are not at the "edge" of their skill level.

You will want to study and progress at your "growing edge"—the skill level where you have a bit of discomfort with new material but where you are not totally overwhelmed. You can expect challenges that trouble you, but they can be overcome.

If you persist and ask lots and lots of questions, you can work through problems that are difficult in the beginning. Everyone is different and finds some problems easy and some problems hard. Finally understanding something you've struggled over can be satisfying. **Paying attention to small details and reviewing to find what you've missed, as well as asking for help when you need it, are keys to success in math.**

Only Smart People Can Do Math

Educators know that the I.Q. score is unreliable for identifying smartness if "smart" is defined as "being a productive member of society who is able to solve the problems presented by life." Using this definition of "smart," Harvard psychologist Howard Gardner identified eight different ways of being intelligent. Only one of these eight intelligences involves mathematics and logic. Chapter 7, entitled "Let's Talk Smarts," elaborates on Gardner's ideas.

Being "smart" (whatever that means) could be helpful in math *but* the most important thing is to *keep trying. Persistent students are successful.* They come to class. They ask questions. They take notes. They ask questions. They do their homework. They ask questions.

I've had many so-called "smart" students who don't do these things. Because they think they "know it all," they miss concepts, become overwhelmed, and drop before the end of class. Often, as class begins, they act as if they already know everything and somehow seem unable to give up that role. Then when they don't understand, they are probably embarrassed to ask questions. Don't be influenced by these students in your classes. Be persistent regardless of how everyone else acts. You will succeed.

Mathematicians Do Math Quickly in Their Heads

The only problems mathematicians complete quickly in their heads are the ones they've done many times before. Recall that there are math problems that have remained unsolved for hundreds of years and no speedy mathematician has solved them.

When mathematicians work at their own level—a level that challenges them, and where they can advance—they use scratch paper—usually lots of it. They write down the problem, the ideas in the problem, what they are trying to do, any thoughts they have, possible solutions, and so on. They draw pictures and diagrams. They look up definitions of words in the problem. They talk to other mathematicians. They start at the end of the problem and try to work backward. They begin again on clean paper to think in new ways. They know that working at this level takes patience and time. They don't give up and are willing to begin again and again until they achieve success.

Brain research supports the idea that speed in mathematics comes from practice, which solidifies the connections and pathways in the brain. The key to speed in math is to slow down first, understand the material, and then practice.

If I Don't Understand Immediately, I Never Will

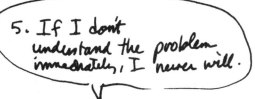

Not true! When you do not understand something, you can say to yourself, "I do not understand *yet.*" That opens the doors in your mind for understanding to creep in.

Creative people know that understanding follows the groundwork of reading, working problems, asking questions, thinking, reworking, and then relaxing for a while. How often have you tried and tried to solve a problem and then, after you finally gave up, had the solution pop into your mind?

One of the most important study skills that you can learn is not to expect to understand immediately! Know that understanding follows working, practicing, reading, asking questions, and living with the ideas for a while.

Some math concepts can be understood immediately. Some of them take mulling over and working with for a while—sometimes for a *long* while. Mathematical historian Lancelot Hogben said math instructors often invite their students to quickly solve math dilemmas that took centuries or more for the human mind to clarify and understand. I can think of math concepts that I finally really understood only as I was discussing them in my math classroom as the teacher. Patience with yourself is important.

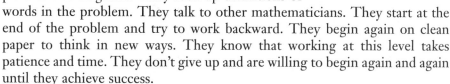

There Is Only One Right Way to Work a Problem

There are many right ways to do every math problem that exists. Actually the right way to first work a problem is the way you understand it best. As you increase your understanding, you can be more open to other ways of looking at it. An open mind that is flexible and willing to experiment with different methods will increase your problem-solving skills.

I Am Too Shy to Ask Questions

This belief about shyness and questions deserves a chapter of its own. Please read Chapter 11, entitled "Questions and All That 'Shy' Stuff."

It Is Bad to Count on My Fingers

Not so! Our number system is based on 10—probably because we have 10 fingers and 10 toes. We have a portable model of our number system with us at all times. When children first learn about numbers, they need concrete models to understand what the abstract ideas mean. Unknowing teachers often skip this concrete model stage.

Many children and adults process information best through their skin and muscles by touching, feeling, and doing rather than by seeing or hearing. These people, called kinesthetic learners (see Chapter 8, "Your Learning Mode"), particularly need the concrete connection with numbers that their fingers provide.

When students freely use fingers in their mathematical calculations, they become more confident and faster in their use. Eventually, most students move beyond this use unless they are in a new or stressful situation where they need grounding.

I teach students how to count on their fingers. Try it. Hold your hands, palms up, so that your fingers are separated and not touching anything. Imagine an electrical charge running through your fingers. Feel numbers such as 3, 8, 3 + 8, 11, etc. The more you practice simple problems like these, the faster you will become.

I often use my fingers to keep focused when I am calculating money in a distracting environment. I also use them to count rests as I play the piano with other musicians in a jazz band. Notice how musicians use their feet, hands, or heads to keep track of time as they perform. Would we dare tell them not to? Unfortunately, in math, using fingers to assist calculations is often foolishly forbidden.

I Should Skip to the Highest Level Math Class that I Can

Placing yourself in a math class beyond your current skill level can cause the Mean Math Blues.

In order to maintain math skills, you must practice them. When you are not using specific math skills, you forget them. They are not gone forever from your brain, but the connections your brain makes to form ideas weaken without use. To reactivate these skills, you need to relearn and practice them.

My personal experience is that these skills often return with more clarity and understanding the second time around. Brain research supports my experience, showing that the relearning process actually forms many new connections in the brain as well as reinforces the old ones.

Since math skills depend on facility with previous skills, in order to advance it may be necessary to repeat a math class. There is no shame in forgetting math skills. It is the biological human condition to forget the skills you are not currently using. **Brain researchers say, "Use it or lose it."**

The math placement test at each college gives good information about the placement best for each student. Repeating previous courses to relearn skills can be a satisfying experience that gets students up to speed and enables them to succeed as they move from course to course.
Enrolling in the next level without relearning the necessary basics can be very frustrating and can produce anxiety about math.

Students who need to extend their math knowledge often find the fastest way is by repeating previous coursework before continuing.

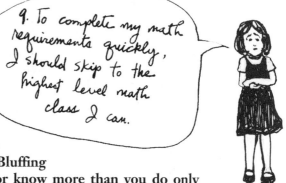

9. To complete my math requirements quickly, I should skip to the highest level math class I can.

Total honesty with yourself regarding math skills is essential to good performance and understanding. Bluffing **yourself into thinking you remember or know more than you do only wastes time in the long run. The more honestly you can admit what you do know and do not know, the lower your level of anxiety will be.**

My Negative Math Memories Will Never Go Away

Brain researchers have discovered that memories are not set in stone—they are fluid and changeable. You revise your memories all the time. For you to grow beyond negativity in math, your new math experiences *must be different* from the old ones.

(*Note:* If you have very vivid and emotionally negative memories of math experiences, you may wish to get some assistance from a counselor to *separate* the negative experience from the math.)

As you gain a broader perspective, you will know more about the boundaries between yourself and your teachers and tutors. You will also know when poor experiences come from avoidance behaviors, lack of information, feeling helpless, or earning poor grades. You will learn when to say: "Excuse me. I believe this negativity belongs to someone else," *or* when to say: "I think I need to act in a different way to achieve success."

To revise negative memories about math, first acknowledge the bad experiences, and look for their causes. It is likely that any insensitive people involved lacked information or were limited by their own math anxiety. Their learning style might have been different from yours and they could not overcome this difference to communicate. As they tried to teach you, they felt responsible for your learning and powerless at the same time. Feeling both responsible and helpless can cause some people to react with anger.

A teacher, parent, tutor, or fellow student may have said or done some very negative things to you, *but* if you now remember and repeat over and over what was said, you are choosing to repeat the negative comments to yourself many times. Recognizing that you repeat the bad experience in your mind can help you take control of these thoughts.

This is the point where you can now intervene and take responsibility for change. You can recognize that those people were wrong and replace those negative messages with new messages that are positive for you.

This book will help you create

- New messages recognizing the difficult circumstances of past times,
- New messages acknowledging how much you have learned since then,
- New messages highlighting your ability to change and to learn now, and
- New messages realizing that you might be surprised at what releasing the old tapes in your mind can accomplish!

REPHRASING THE MYTHS CAN REFRAME THE EXPERIENCE

Read the following statements and check those that you would now be willing to accept as true.

❑ 1. I can do some math. Some math I can't do yet. Some math I need never do.

❑ 2. There are both easy and hard math problems for everyone.

❑ 3. Persistence in math helps more than smarts. Knowing how I am smart can help me to learn math.

☐ 4. Mathematicians can quickly do problems that they have practiced. When they learn new math ideas, they work very slowly and use lots of scratch paper.

☐ 5. Understanding math takes time.

☐ 6. There are many right ways to work and think about each math problem.

☐ 7. Feeling shy or reluctant to ask questions is common. I can seek out safe situations to ask what I need to know.

☐ 8. Using my fingers to calculate can be quick and helpful. In fact, I need to use whatever works for me.

☐ 9. By taking math courses at my skill level I will complete my math requirements more quickly than if I jump in over my head.

☐10. Negative math memories can fade and be replaced in time.

Pushing Your Limits

1. Write any beliefs from the True/False quiz that you marked True. What was your reaction when you found they are false? Write down new information that helps you disagree with your old beliefs. Did I make a good enough case to convince you to change your mind? Discuss these thoughts with instructors and other students. If you find a particular "negative math thought" that you still believe true, give extra attention to that issue. Write down a "pretend" argument with me. Write what you would say to me about the thought and then write the response that you think I would say to you. Keep considering the thought and my points. Don't give up your belief until you are convinced.

2. Experiment with this exercise to change a negative math memory. It is adapted from Richard Bandler's Neuro-Linguistic Programming Model for change.

First, think of a situation (not a math situation) that was very positive for you in some way—a time when, for that moment, you felt "in charge"—in charge of yourself and your reaction to what was happening—an experience where you felt competent and knew that you could handle anything that happened, if even just for the moment—an experience of "well being" when you felt peaceful or energized.

You may wish to close your eyes. Breathe deeply and sharpen your mind. Acknowledge that it is O.K. to be human and that you don't have to be perfect in order to be competent. For now, you are adequate for this situation you are remembering, and "what and who you are" is enough. Imagine your expression and what you see/hear/feel/know. Brighten the image, turn up the sound, and increase the feelings.

Now run a video in your mind of your negative math experience. What does the competent, adequate person of your positive experience notice? Write down what you would say to yourself and to each of the people in your video. Write down how you would edit that experience to make it positive. Rewrite that video script in your mind to change it. You can call in imaginary or real consultants to help you.

3. Choose one of these goals for your math class this week and evaluate your progress one week from today:

- Relax consciously in class by breathing deeply.
- Start or attend a study group.
- Bring problems to class with questions.

Record your goal and results on the "Goals Page" of your journal.

4. Choose one of these actions to get feedback and increase your math understanding. At the end of the week, evaluate whether this was helpful or not. Record your choice and evaluation.

- Solve a problem several different ways.
- Visit the teacher during office hours.
- Draw a picture of a problem.

5. A new feature—Mastering Math's Mysteries—begins in this chapter. Its purpose is to give you math work to try out the techniques you learn in each chapter. Remember that anxiety lowers when you take charge rather than avoid. Try Mastering Math's Mysteries, Chapter 3, to practice counting and using your fingers with math.

PREVIEW: MASTERING MATH'S MYSTERIES

To feel good about math, recognize that even the best mathematicians do not recognize and understand everything mathematical that they see. They don't expect to. Work on your expectations also. If you expect to understand math immediately, you set yourself up for disappointment. The Mastering Math's Mysteries in each of the following chapters are designed to expose you to math ideas so that you can practice the techniques you are learning in this book. Because math students at all levels get the blues and will be reading this book, it has been difficult for me to choose math topics to introduce here. Try the Mysteries, but if you are not ready for the ideas presented, that is O.K. It means that I have not chosen the math that you can learn at your level. Enrolling in a basic math course or working through a book such as *The Only Math Book You'll Ever Need* by Stanley Kogelman and Barbara R. Heller will provide math material at an appropriate level for you.

More difficult and challenging math exercises are included in the Appendix for anyone willing to try them. If you are currently taking a math class, you may have plenty of mathematical ideas to practice and may wish to read only the chapters themselves, practicing the techniques on your homework.

Start slowly. Math is not a competition. Take all the time you need with each topic. The solutions to the exercises are in the back of the book.

The following topics are included:

Chapter 3, Counting
Chapter 4, Patterns and Square Numbers
Chapter 5, Triangular Numbers
Chapter 6, Fibonacci Numbers
Chapter 7, Introduction to Fractions: Adding & Subtracting
Chapter 8, Working with Halves
Chapter 9, More on Adding & Subtracting Fractions
Chapter 10, Multiplying Fractions
Chapter 11, Dividing Fractions
Chapter 12, Sum, Difference, Product, Quotient
Chapter 13, Memory Devices for Fractions
Chapter 14, Webbing
Chapter 15, Spatial Visualization
Chapter 16, Problem Solving
Chapter 17, Dress Rehearsal Test

Counting

You do math when you drive, spend money, and *count*. Depending on your life experience, your math skills differ from others'. Practice a few of your skills now. Check your answers with the solutions in the Appendix.

1. Write the counting numbers from zero to nine.

_____, _____, _____, _____, _____, _____, _____, _____, _____, _____

The 10 symbols that you wrote above are the basis of our number system. There are probably 10 symbols because we have 10 fingers and 10 toes. (A bit of trivia: If we had eight fingers and eight toes, our number system would probably only have eight symbols, leaving out the 8 and the 9. Counting with eight symbols could look like this: 0, 1, 2, 3, 4, 5, 6, 7, 10, 11, 12, 13, 14, 15, 16, 17, 20, 21, . . .)

2. Sometimes we count by twos like this: 2, 4, 6, 8, 10, 12, 14, . . . Write the next three numbers of this sequence after 14: ____, ____, ____

3. Count by threes from 3 to 30. Complete the sequence below:

_3_, _____, _____, _____, _____, _18_, _____, _____, _____, _30_

The sequence that you wrote above could help you remember multiplication by three. Here's how: Say the sequence aloud using your fingers. Raise one finger with each number. Later, if you do not immediately remember 3 times 6, count by threes on your fingers up to the sixth finger, which gives you the answer 18.

4. Count by fours from 4 to 40.

_4_, _____, _____, _____, _____, _____, _28_, _____, _____, _40_

Do you have trouble multiplying by four? Practice saying the sequence above aloud using your fingers. Raise another finger with each number to cement these numbers in your mind.

5. Count by fives from 5 to 50. Write the numbers here.

_5_, _____, _____, _____, _____, _____, _35_, _____, _____, _____

6. Count by sixes from 6 to 60. Write the numbers here.

_6_, _____, _____, _____, _30_, _____, _____, _____, _54_, _____

PART TWO

Change Your Thoughts

MANAGING THE MEAN MATH BLUES

4

Thoughts In Charge

"I've been absolutely terrified every moment of my life—and I've never let it keep me from doing a single thing I wanted to do."

GEORGIA O'KEEFE

HOW HUMAN BEINGS WORK

Your thoughts, emotions, behaviors, and body sensations—different and distinct—are interrelated. Each one influences and is influenced by the others, as shown in the Interrelationship Chart below (Greenberger & Padesky, 1995).

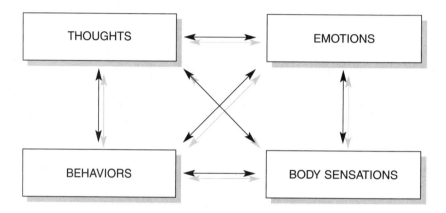

- **Thoughts** include your beliefs, ideas, opinions, knowledge, rules, and experiences.
- **Emotions** are feelings and moods.
- **Behaviors** are how you act—what you do.
- **Body sensations** are the physical reactions of your body.

Stanford neuroscientist Robert Sapolsky's research shows that "humans can activate stress responses by thoughts." In this chapter and the next two chapters, you will learn tools to change your thoughts and deactivate your stress responses that occur in the form of overwhelming negative feelings and debilitating physical symptoms.

Compare the two contrasting interrelationship charts shown in Figures 4.1 and 4.2. See how the four aspects "cause and effect" each other. Specifically, start by reading the "thoughts" and notice how they influence the rest. Notice that different thoughts change the other three components.

It is through thoughts and behaviors that you can consciously intervene in these interrelationships to change the other outcomes. Chapters 5 and 6 will teach you how. For now, the exercises on the next few pages will identify where you are and where you wish to be.

CHART YOURSELF

At this point, you may believe that you are powerless to change your thoughts. This is not true. Recognition brings consciousness and facilitates change.

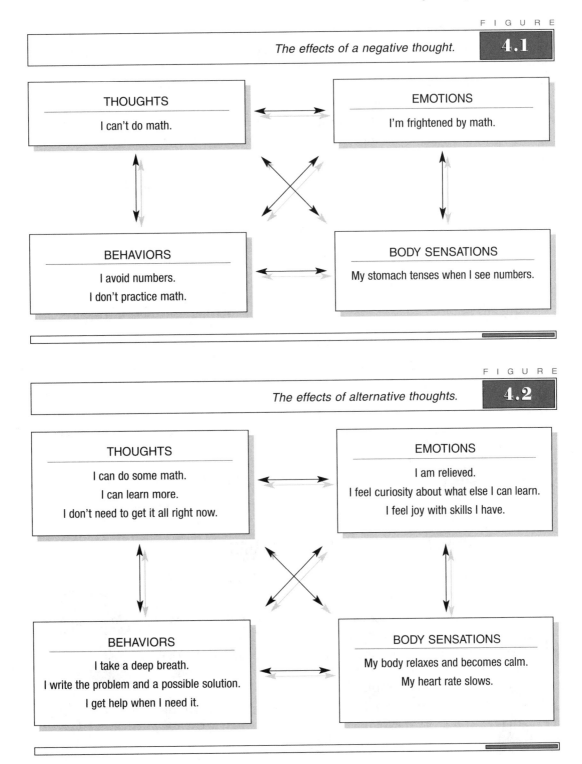

F I G U R E

4.1

The effects of a negative thought.

THOUGHTS

I can't do math.

EMOTIONS

I'm frightened by math.

BEHAVIORS

I avoid numbers.

I don't practice math.

BODY SENSATIONS

My stomach tenses when I see numbers.

F I G U R E

4.2

The effects of alternative thoughts.

THOUGHTS

I can do some math.

I can learn more.

I don't need to get it all right now.

EMOTIONS

I am relieved.

I feel curiosity about what else I can learn.

I feel joy with skills I have.

BEHAVIORS

I take a deep breath.

I write the problem and a possible solution.

I get help when I need it.

BODY SENSATIONS

My body relaxes and becomes calm.

My heart rate slows.

- Copy the chart below on a separate piece of paper or in your journal.
- First write a specific thought about math under "Thoughts." You could use your strongest negative belief from the last chapter.
- Next fill in the related emotions, behaviors, and body sensations in the appropriate places.
- Make a second chart using the rephrasing of your negative thought. Compare the two charts. Notice how changing the thought affects the rest.

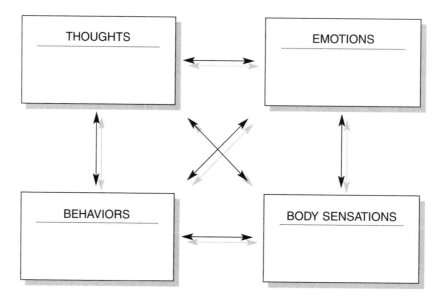

> *"Our thought is the key which unlocks the doors of the world."*

SIGNS OF THE MEAN MATH BLUES

Figure 4.3 lists the symptoms of the Mean Math Blues from Chapter 1, and others.

- Circle the three signs in each category that you have most strongly experienced.
- Next, choose one thought that you circled. Does it relate to any of your circled behaviors, emotions, or body sensations?

- Make an Interrelationship Chart for each circled thought. What do you notice? Where might you be able to make a change consciously? The next two chapters will help you to do this.

Symptoms of the Mean Math Blues.   **4.3**

THOUGHTS	
I am helpless.	I'm alone with math.
I am stuck.	I can't do math.
I am bored.	I am inadequate.
I'm out of control.	I've never done math.
I will fail.	I will never do well.
I don't trust math.	Math teachers hate me.
I am incompetent.	Old feelings don't
I should do better.	change.

EMOTIONS	
Guilty	Depressed
Helpless	Frustrated
Terrified	Bored
Anxious	Panicky
Foolish	Angry
Nervous	Shameful
Despairing	Annoyed
Embarrassed	Hopeless
Discouraged	

BEHAVIORS	
Avoid math.	Tune out in class.
Act confused.	Don't take notes.
Avoid class.	Don't ask questions.
Sit in the back.	Refuse to listen.
Sleep in class.	Avoid homework.
Act bored.	Come late to class.
Complain.	Talk negatively to myself.
Waste time.	Blame the teacher.
Act helpless.	Skip study groups.

BODY SENSATIONS	
Nausea	Palpitations
Blurred vision	Hives
Shaking	Headaches
Crying	Backaches
Tension	Stomach ache
Sweating	Dizziness
Insomnia	Fatigue
Racing heart	Agitation
Lethargy	Stress

WISH LISTS

Review Figure 4.4. You may not relate to many of the entries in these lists *yet*. Use your imagination.

- Read these lists and boldly check those that you have already experienced.
- Circle three from each list that you would like to experience.
- Next, for each thought that you circled, make an Interrelationship Chart.

FIGURE

4.4   *Wish lists to manage the Mean Math Blues.*

THOUGHTS	
Practice helps.	I can ask questions.
I have skills.	I have learned before.
I am smart.	I overcome challenges.
I am creative.	I choose to be a good student.
I can get support.	I have resources.
I do my best.	I intend to understand.
I learn well.	Understanding takes time.
I can do math.	My self-worth does not
I am worthwhile.	depend on math.

EMOTIONS	
Joyful	Excited
Ecstatic	In control
Calm	Satisfied
Curious	Grateful
Astonished	Delighted
Convinced	Cheerful
Contented	Complete
Balanced	Capable
Proud	

BEHAVIORS	
Ask questions.	Write and monitor my
Take breaks.	thoughts.
Exercise.	Come to class prepared.
Sit in front.	Consult the teacher.
Persist.	Write and take dress
List my resources.	rehearsal tests.
Do my homework.	Consider alternatives.
Encourage others.	Check my answers.
Breathe and relax.	Savor my good work.
Encourage myself.	Form a study group.

BODY SENSATIONS	
Relaxation	Calmness
Being refreshed	Steady heartbeat
Alertness	Strength
Warmth	Health
Robustness	Being well rested
Serenity	Peace
Being coolheaded	Being composed
Hardiness	Being clear-minded
Attentiveness	

ADVICE FOR STRUGGLING MATH STUDENTS

I interviewed **creative, non-math professionals** about how they think about math. They told me to give you this advice.

> This sounds nuts—my best advice to struggling math students is to quit struggling. I wish I knew then what I know now—which is to relax, just in all areas. When I relax, I find that I'm much more available to get it, to learn it.
>
> But I was so convinced that I couldn't do it. While I was being taught, I would spend time in my own mind telling myself that I couldn't do it. I was so

uptight about not being able to do it that I didn't know how to stop struggling against it. And given the chance to struggle a little, I struggled a lot.

The best thing that I could have done would be to just quit struggling against it. And let the grades go.

—Giovanna Piazza, Priest and Ethics instructor

Take all the barriers you have about math—all the preconceived notions that you have about math as a subject—put them in a little box and burn it or bury it.

Then relax with math and let it infuse you instead of worrying about what you're going to do with it. Just let it come in to you. If you're doing math, look at it and notice the relationships and places where you will use the concept. You can make up a formula once you have a concept. I didn't know that. Once you have the pattern then you can do not only math but many things.

—Bobbi Nesheim, Ph.D., psychotherapist and owner of Center for Creative Growth

Pushing Your Limits

4

CHAPTER

1. In your journal, write the three strongest thoughts, emotions, behaviors, and body sensations that you circled in the Signs of the Mean Math Blues list. Make Interrelationship Charts using your strongest thoughts. Is there a relationship between those thoughts and your strongest emotions, behaviors, or body sensations? What effect would changing any of the thoughts or behaviors have on the thoughts, emotions, behaviors, and body sensations?

2. Record the three thoughts, emotions, behaviors, and body sensations that you circled in the Wish Lists. Choose two thoughts you find most desirable and make Interrelationship Charts for them. Are the emotions, behaviors, and body sensations that you circled related to those thoughts? Write about this.

3. As you work through this book, return to the Wish Lists of thoughts, emotions, behaviors, and body sensations. As you learn to manage the Mean Math Blues, can you boldly check more that you have experienced? Notice how your actions (behaviors) can influence change.

4. Experiment with the square numbers in Mastering Math's Mysteries, Chapter 4. Record both positive and negative thoughts that run automatically through your mind as you answer these questions. If you don't see the patterns right away, come back to them later. Approach these pages with curiosity.

Patterns and Square Numbers

LOOK FOR PATTERNS. Just noticing patterns in the world around you develops math skills. Train yourself to look for patterns in license plates, money transactions, telephone numbers, social security numbers, and whenever you see numbers. Notice geometrical patterns too, such as the shape of lights, clocks, walls, fences, buildings, fountains, trees, and flowers.

In the last chapter, you looked at counting patterns. Counting by ones: 0, 1, 2, 3, 4 . . . Counting by twos: 2, 4, 6, 8, 10 . . . Counting by threes: 3, 6, 9, 12, 15 . . . Practice those sequences now, using your fingers.

SQUARE NUMBERS. Check out this pattern: 1, 4, 9, 16, 25, 36, . . . are called the square numbers. Also known as perfect squares, each square number is formed by multiplying one of the counting numbers (1, 2, 3, 4, 5, 6 . . .) by itself. This process is called "squaring."

The square number 1 is 1 • 1. (The raised dot • is a symbol for multiplication.) The square number 4 is 2 • 2. The square number 9 is 3 • 3. The square number 16 is 4 • 4, and so forth.

1. Fill in the blanks below by noticing the patterns in each row and column. (Remember, the solutions are in the back of the book.)

1•1	2•2	3•3	4•4			7•7			10•10		
1	4	9			36	49	64	81		121	144

2. In the blanks below, write the square numbers that are represented by the squares underneath. Notice that each square number is the number of small squares inside the large square. Notice the lengths of the sides. For example, the square that is 3 units on each side has 9 small squares inside because 3 • 3 is 9. Draw the next two squares.

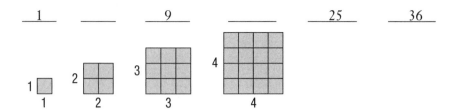

| __1__ | _____ | __9__ | _____ | __25__ | __36__ |

3. Make a list of the "perfect squares" from 1 to 100. There are exactly 10 of them.

									100

4. If you have a calculator, continue your list of perfect squares up to 400. Hint: 400 is the last perfect square in your list and comes from squaring 20.

121									400
11•11	12•12	13•13	14•14						20•20

5. The meaning of 1^2 is $1 \bullet 1$. In other words, 1^2 means "multiply 1 times itself" to get 1. The small, raised "2" in 1^2 is called the **exponent.** The "1" in 1^2 is called the **base.** When the exponent is a 2, we say that we are "squaring the base."

When we change the base, as in 2^2, we multiply the new base 2 by itself to get 4. We say that we are squaring 2.

The symbol 1^2 is read "one squared" or "one to the second power." 2^2 is read "two squared" or "2 to the second power."

Whenever you write a counting number with a 2 in the exponent, you are "squaring" the counting number and will get a "square number" for an answer. Notice the patterns to fill in the empty blanks below.

1^2	2^2	3^2	4^2			7^2			10^2
1	4		16	25	36	49	64	81	

5

Neutralize Negative Math Thoughts

RALPH WALDO EMERSON

"The ancestor of every action is a thought."

THOUGHT DISTORTIONS

Repetitive negative thoughts about mathematics contain distortions that warp your thinking (Beck & Emory, 1985). This chapter defines seven Thought Distortions (Burns, 1999) in negative thinking and asks you to identify them in math situations. The more effort you put into this chapter, the more aware you will become of your negative thinking. As you recognize negative math thoughts, identify Thought Distortions, and use Intervention Strategies (Burns) from the next chapter, you will learn to consciously change your negative feelings, behaviors, and body sensations. The 10 false beliefs from Chapter 3 all contained Thought Distortions.

THOUGHT DISTORTIONS
in Negative Thinking

Read through these definitions and their examples. Copy the Thought Distortions into your journal, and make notes and symbols of meanings. Then place a check beside those examples you have thought. Write more examples of your own. Finally, practice identifying Thought Distortions and neutralizing negative thoughts in the three math situations in this chapter.

1. Absolute Thinking

Absolute thinking recognizes only two alternatives, and they are opposite extremes—*black or white—all or nothing.* According to this thinking, on a scale from 1 to 10, everything is either a "1" or a "10." Black or white thinking ignores the alternatives—the gray, in-between areas. Very few issues are *either/or.* Watch for: **1 or 10, Black or White, Good or Bad, All or Nothing.**

EXAMPLES

❑ Either I get an "A" grade or I am a failure. B's and C's are unacceptable.

❑ Either I get these three math problems or I can't finish and understand my assignment.

❑ Math is always hard.

❑ Only smart people can do math.

❑ If I don't do well on this homework, I won't pass my class, finish my degree, or ever get a job.

2. Overgeneralization

Overgeneralizing makes broad conclusions, often dire predictions, about the future based on one event. When statisticians make generalizations, they ask *many* people or view *many* situations first. When you make sweeping life decisions based on one incident, you are overreacting.

EXAMPLES

❑ My whole life is ruined because of this problem.

❑ If I can't complete this homework, I will never be able to get math.

❑ Since I failed the test, I know I will always do poorly in math.

❑ I can't do math.

❑ I used a math tutor once. He was critical and unhelpful. I don't work well with math tutors.

❑ My high school math teacher was awful. Math teachers are insensitive.

❑ I am a math failure because I don't understand Uniform Spaces.

3. Mind Reading

In *mind reading* you imagine that you know what someone else is thinking or feeling. This causes you to draw inaccurate conclusions and behave based on a fantasy.

EXAMPLES

- ❏ I think the teacher doesn't want to help me, so I won't ask.
- ❏ The others in the class think I'm a math failure. They wouldn't like me to ask them a question, so I won't.
- ❏ Other students think I'm stupid when I ask questions, so I will stop.
- ❏ Mathematicians always do problems quickly in their heads. I should understand these math problems right away.
- ❏ I better not ask questions because the teacher and students will know that I'm inadequate.

4. NEGATIVEpositive

NEGATIVEpositive filters out and minimizes *positive* input while focusing on the negative. Obsessing on the few negative things that happen while ignoring the rest is a common human reaction. However, it is an inaccurate perception of the world. Amplifying the negative is so overwhelming that most people deny and avoid rather than using the negative information as feedback for learning and making changes.

EXAMPLES

- ❏ My glass is half empty.
- ❏ I look at my 75% test grade and only notice what I got wrong, not the fact that I got many things correct.
- ❏ I say something I regret in class and keep berating myself for not being perfect.
- ❏ My teacher says I do well with problem solving. I think she's just being nice. I know that I missed that word problem last week and got stuck on the one in class, so she's wrong.
- ❏ I got an "A" on my first three exams. Now I got a "C." I knew I couldn't do math.
- ❏ I cannot do a math problem so I think I cannot do the whole homework assignment and probably cannot do any math from here on.
- ❏ I work with a study group and am uncomfortable about one incident during the four-hour meeting. The next day I have maximized this negativity to the entire session and vow to never attend one again.

5. Crystal Ball Thinking

Crystal ball thinking predicts the future and colors it negatively. Predicting the future ignores the fact that you change every day of your life. Your life is a process of change. You are not a static being. Unfortunately, you can actually make bad things occur in your life by believing that the worst will happen, and then behaving as if this were true.

EXAMPLES

❏ I *will* never be able to do math.

❏ I *will* never graduate.

❏ If I don't understand a problem quickly, I never *will*.

❏ My memory of my negative math experience *will* never go away.

❏ I won't pass that math test *next week*.

6. Feeling = Being

Feeling = Being equates the *feelings* you have about yourself or your moods with who you truly are as a person. You forget that you are not your feelings. Your feelings are the results of your thoughts and behaviors and are separate. Your feelings are part of you, but they are not accurate pictures of your whole self. **The Truth: Feeling ≠ Being.**

EXAMPLES

❏ I feel anxious about math or math tests, so I think I am unable to do well in math or will perform poorly on math tests.

❏ Because I feel inadequate, I believe I am inadequate.

❏ I am too shy to ask questions. (See Chapter 11.)

❏ I feel overwhelmed, so I think I can't learn this.

❏ I feel bad or dumb or guilty, so I think I am bad, dumb, or guilty.

7. Shoulding

Shoulding uses words like "should," "must," "have to," and "ought." These words imply a negative, parent-like order. Most people don't accept orders well. These words attempt to place blame. The power shifts from internal to external. You may feel coerced and perhaps a bit rebellious when spoken to in this manner—whether by yourself or someone else. Choice and goodwill shrink under orders. **Watch for: I should . . ., I must . . ., I have to . . ., I ought to . . .**

EXAMPLES

❏ I should do all my homework.   ❏ I shouldn't make mistakes.
❏ I must get very good grades.   ❏ I ought to be perfect.
❏ I have to be good.   ❏ I should be the best.
❏ I ought to do this.   ❏ I shouldn't count on my fingers.

You can make many thoughts more reasonable and truthful by replacing the words above with **"I choose to"** or **"I want to"** or **"I will."**

- I choose to do all my homework.
- I want to get very good grades.
- I want to be good.
- I choose to do this.

- I will make mistakes. (It's human.)
- I want to be perfect. (It's not human.)
- I would like to be the best.
- I choose to use my fingers when I need to.

EDITING NEGATIVE MATH THOUGHTS

Now you can practice identifying Thought Distortions and rewriting negative math thoughts as neutralized thoughts without Thought Distortions.

DIRECTIONS

- Read the following situations and the related automatic negative thoughts.
- Identify the Thought Distortions. You may find more distortions than I have.
- Next write down some alternative neutral thoughts that could relate to the situation before you read my ideas. Make the alternative thoughts true statements.

Situation 1: Homework

You are doing well with your math homework but get stuck on several problems in a row.

You might automatically think some of the negative thoughts illustrated in Figure 5.1. Before reading on, identify the Thought Distortions contained in each of these negative thoughts.

(a) _____

(b) _____

(c) _____

(d) _____

(e) _____

Negative thoughts regarding homework.

These are the Thought Distortions I identify. You might find more:

(a) Crystal Ball, Absolute Thinking, NEGATIVEpositive
(b) Crystal Ball, Absolute Thinking, Overgeneralization
(c) Feeling = Being
(d) NEGATIVEpositive, Overgeneralization
(e) Crystal Ball, Overgeneralization, Mind Reading

Before reading on, rephrase each negative thought into a more neutral and useful statement. Make it a true statement to be effective. Then read my suggestions, illustrated in Figure 5.2.

ALTERNATIVE RELATED THOUGHTS

F I G U R E

5.2   *Alternative, useful thoughts regarding homework.*

The learning process is challenging.
There must be something
I don't quite understand here.

This has happened to me before.
I have worked through it.

Maybe I need to ask some questions,
or to do some of the examples again.

I have many resources to assist me—the book,
notes, examples, the instructor, friends, and tutors.

I can use the answer from the answer key
and try to work backward to get the process.

Just because these few problems are difficult
doesn't mean all the rest will be difficult too.
This is an opportunity for me to figure out
what I misunderstood and correct it.

I could skip ahead and see if there are some problems there that
I can do. I will mark these problems so I can come back to them.

Situation 2: Math Class

You are in your math class. The teacher is working problems. You pretty much understand what she is doing and
then suddenly you are totally lost. The symbols begin to look
like Greek—and you don't speak Greek.

You might automatically think some of the negative thoughts shown in Figure 5.3.

Negative thoughts regarding math class.

a) Everyone understands what to do except for me.

b) Because I don't understand this, I will never be able to do math.

c) I feel like an idiot.

d) I should understand this.

e) The teacher will be upset with me if I ask questions about this.

Before reading on, identify the Thought Distortions contained in each of these negative thoughts.

(a) _____

(b) _____

(c) _____

(d) _____

(e) _____

These are the Thought Distortions I identify. You might find more:

(a) Mind Reading, NEGATIVEpositive

    (b) Overgeneralization, NEGATIVEpositive, Absolute Thinking

    (c) Feeling = Being

    (d) Shoulding, Overgeneralizing

    (e) Mind Reading

    Before reading on, rephrase each negative thought into a more neutral and useful statement. Make it a true statement to be effective. Then read my suggestions, illustrated in Figure 5.4.

F I G U R E

5.4

Alternative, useful thoughts regarding math class.

ALTERNATE RELATED THOUGHTS

_____

_____

_____

_____

_____

FIGURE

Continued.   **5.4**

I am just having a normal human reaction
to seeing something new and strange
for the first time.

When I do not understand,
I hang in until I do understand.
I have come so far in my math learning.
Here is another challenge.
This is all it is.

I can act positively by taking a deep breath
and congratulating myself for being courageous
enough to put myself in this class.

Establishing a relationship with my teacher might
make asking questions during, or after, class easier.

I will take notes or ask a question
so I can clarify this concept.

My mind depends on my body.
Perhaps I need to get more sleep or more exercise.
Perhaps I need to eat more nutritiously before class
or drink more water to help my brain function optimally.

Situation 3: Consulting the Teacher

You finally get the courage to go to the teacher's office to ask questions. The teacher looks at your math work and says, "You should know this."

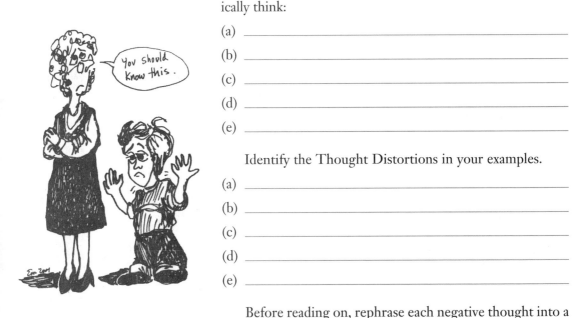

Write down negative thoughts some students might automatically think:

(a) _____

(b) _____

(c) _____

(d) _____

(e) _____

Identify the Thought Distortions in your examples.

(a) _____

(b) _____

(c) _____

(d) _____

(e) _____

Before reading on, rephrase each negative thought into a more neutral and useful statement. Make it a true statement to be effective.

ALTERNATE RELATED THOUGHTS

(a) _____

_____

(b) _____

_____

(c) _____

_____

(d) _____

_____

(e) _____

_____

Here are my suggestions:

- "Maybe I should know this *and* I don't. It is O.K. to not know something. This is a place for me to start improving my understanding."

- "I have come to the right place to find out how to further my understanding and learn what I 'should' know in order to advance in this course. If I am in the wrong course, it is best to find out now and get into the right course to lower my frustration."

- "I can say to my teacher, 'Yes, you're right. I should know this and I don't. Would you help me figure out how to best learn this and get my questions answered? I care about learning this material and passing this course. I appreciate your time and assistance. Do you have any further recommendations for places or ways to get assistance? This is what I am able to do now. Perhaps there are examples in the book you could recommend or perhaps you could help me identify the error in my thinking that is causing me so much difficulty with this concept.'"

Pushing Your Limits

CHAPTER 5

1. Copy the Thought Distortions to carry with you in your math book. List them in your journal. Use them to examine automatic negative math thoughts whenever they occur. Become more conscious of your own cognitive processes to prevent negative thoughts running loose in your mind.

2. List your negative math thoughts in your journal. Identify the Thought Distortions. Identify the negative math thoughts that you most wish to change. Write down alternative neutral thought substitutions. Try some of the interventions listed in the next chapter. If you need assistance or an outside opinion, ask your instructor or a supportive friend for fresh ideas. Sometimes we are too close to ourselves to recognize objectively what is going on.

3. Analyze this situation: You work really hard on a math problem and can't do it. When you go to class, you mention that you didn't get it. You hear another student say, "That was easy." List possible automatic negative thoughts and their Thought Distortions. Which of the following alternative related thoughts seem reasonable to you?

- "Most things are easy once I understand and practice them."

- "That other student is expressing relief or enthusiasm that she understands or she might be bragging and insensitive. I do not have to hear her comment as criticism of me. My responses are my responsibility. I can't change the world—only my reaction to it."

- "Right now I don't understand this problem and I, being human, have a right to the learning process and its challenges. People do not all learn at the same rate and that's O.K. I am not in competition with the others in my class. Learning math is not a race."

4. Choose one of these goals to work on this week in math and evaluate your progress one week from today:

- Recognize how much more you know now than you did two weeks ago.
- Summarize what you learned in class today.
- Turn in completed homework on time.

Continue to record these goals and your evaluation on the "Goals Page" of your journal.

5. Choose one of these actions to get feedback and increase your math understanding. Evaluate and record the usefulness of this action at the end of the week.

- Ask yourself if your answers make sense.
- Work the examples from class over until you can do them without consulting your notes.
- Consult your study group on a problem.

6. Mastering Math's Mysteries, Chapter 5, gives you more practice with patterns—a great math skill to cultivate. What do you think about when you hear the words "Triangular Numbers"? Draw your ideas and then work through this chapter's mysteries to see what I am calling "Triangular Numbers." As you work these exercises, notice what thoughts you are having. Are you having any automatic negative thoughts? What Thought Distortions do they contain? Can you reframe any of these thoughts to neutralize them?

Triangular Numbers

The "Triangular Numbers" are another interesting sequence of numbers. They are:

1, 3, 6, 10, 15, 21, 28, 36, 45, 55 . . .

These numbers earn their name because they form the patterns shown here:

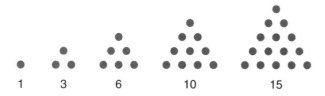

1. Count the dots in each triangle to get the Triangular Number. Notice that dots are in rows, and each row has one more dot than the row above. Fill in the table below for the Triangular Numbers 10 and 15. Look at their triangles above for help.

TRIANGULAR NUMBER	ADD DOTS IN ROWS:
1	1
3	1 + 2 = 3
6	1 + 2 + 3 = 6
10	
15	

2. Notice how each new triangle is the same as the preceding triangle but with a new row that is one dot longer. How many new dots would there

be in the triangle for the next Triangular Number, 21? _____ Draw the triangle for 21.

3. Compare each Triangular Number to the next. Do you see a pattern in the numbers themselves or triangles above that would allow you to find all the Triangular Numbers without drawing their triangles? (Hint: Look at the differences between the numbers as they increase, *or* notice the rows of the triangles and think about adding.) Write down your pattern, along with the next four Triangular Numbers after 15.

<u> 1 </u> , <u> 3 </u> , <u> 6 </u> , <u> 10 </u> , <u> 15 </u> , <u>   </u> , <u>   </u> , <u>   </u> , <u>   </u> , <u> 55... </u>
 add 2   add 3   add 4   add ?   add 6   add 7   add 8   add 9   add 10   add ?

4. How do the figures shown below illustrate the Triangular Numbers? Instead of counting dots, count boxes. Notice that the added column gets one box higher each time, forming higher and higher staircases. Draw the next two figures for Triangular Numbers 15 and 21. Count the number of boxes in each figure that you draw to be certain that you have 15 and 21.

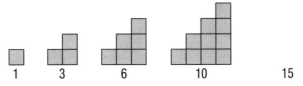

   1   3   6   10   15   21

Notice how much you add each time to get the next number. Describe the pattern. (See #1.)

_____

_____

_____

_____

6

Intervention Strategies for Negative Thoughts

"All things change—

so can we."

JULIA CAMERON

In their book *Mind Over Mood*, clinical psychologists Dennis Greenberger and Christine Padesky say, "People do not usually overcome anxiety until changes in thoughts are accompanied by changes in avoidance behaviors."

You have now identified thoughts that you wish changed, and you are ready to change the behaviors that go with them. Burns (1999) identifies many interventions you can make. The figure below shows eight. Check the ones that appeal to you now as possibilities. Some of these strategies might need to "grow on" you a while to become acceptable. Return to this section and reread the possible intervention strategies from time to time. You may discover more strategies that will be useful.

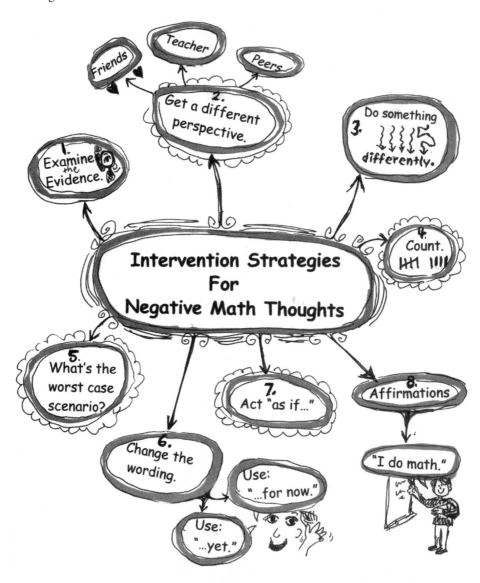

INTERVENTION STRATEGIES FOR NEGATIVE MATH THOUGHTS

1. Examine the Evidence

What is the evidence that your negative thought is really true? Are you over-reacting? What is the evidence that the thought is false? What would you do differently if this thought were false?

EXAMPLE: You think you will fail math.

- Have you truly gotten failing test scores?
- Are the low grades the result of your neglecting studying and homework?
- Have you refused to get assistance or ask questions?
- What is the evidence that you are, in fact, failing?

2. Get a Different Perspective

EXAMPLES

- Tell yourself what you would tell a *close friend* who has this thought. Would you want him to be handicapped by this negative thought? We are often much harder on ourselves than we are on others. Could you not choose to be as kind to yourself as you would to your friends?
- Speak to yourself or write down what a *good friend* would say to you about this negative thought. A close friend is probably more objective and positive than you would be for yourself.
- You may feel like you are the only one struggling. Talk to your teacher, your tutor, and other students in the class to see how realistic your thought is.

3. Do Something Differently

Behave in a new way to get a different result. Identify actions that contribute to your negative math thoughts and learn from what you have recognized. Change these actions to some behaviors that are new to you.

EXAMPLES

- Recognize that you cannot expect yourself to understand math when you do not practice by doing your homework
- Recognize that you may not understand the class lectures because you don't ask enough questions or take enough notes or get enough sleep to stay alert.

4. Note the Number

Count and record the number of times that you think these negative math thoughts. Recognition is one of the first keys to bringing them to consciousness and changing them.

5. Identify the "Worst Case Scenario"

Ask yourself, "What is the worst thing that can happen in this situation?" Often the worst thing that can happen isn't so bad after all. You can survive all kinds of "terrible" things. Sometimes it is the fear that is worse than the consequence.

EXAMPLE

You have a question in your math class and are reluctant to ask it. The worst thing that could happen might be that the teacher will yell or that the teacher will refuse to answer or that someone might laugh. None of those three events is life threatening. Each possibility shows insensitivity on someone else's part—not an indication that you are wrong for asking your question. Sometimes we are willing to risk possible consequences when we realize how unlikely they are to occur.

6. Change the Wording

Restate the thought in a way that is neutral or could actually be positive. Add the words "right now," "for now," or "yet."

EXAMPLES

- Change: "I will fail math" to "Right now I cannot predict the future and I can certainly do some things to prevent failure."
- Change: "I can't do math" to "Right now I am unable to do these math problems."
- Change: "I don't understand" to "I don't understand *yet.*"
- Change: "I'm not prepared" to "I'm not prepared *yet.*"
- Change: "I'm never going to get this" to "I get it up to this point."

7. Act "As If"

Act as if you had whatever trait you lack or are whatever you would like to be. Ask yourself, "How would I look? What would I hear differently? What would I say? How would I behave? Assume new thoughts and behaviors—don't just pretend. "Try on" success.

EXAMPLES

- If you want to be a successful math student, consider how good math students act. What behaviors do they do? They go to class, ask questions, complete homework, work with classmates and teachers, read the textbook, use tutoring services, admit they don't understand . . . What else would a successful student do?

- Public speakers frequently act "as if" they feel confident. They put their shoulders back, smile, speak up, make eye contact, and act "as if" they know what they're talking about. As in the song Anna sings in the musical *The King and I*, they fool themselves as well.

8. Affirm Your Best

Read the "Math Affirmations" in the following box aloud to create a mental math picture that is supportive, hopeful, and strengthening. Coupled with asking questions and working math problems, these affirmations open your subconscious mind to confidence with math and to positive feelings about math, if you repeat them often.

Math Affirmations

I do math.

- My mind is open and I invite math in. It is not a problem when I don't understand right away. With continued practice and many questions, my understanding increases. My goal is to progress—to be able to do a little more, step by step, each day. Understanding follows. It is O.K. to say, "I do not yet understand." I am willing to learn.

- I look back to notice how far I've come and I enjoy my progress. I know more than last week and much more than a month ago. Small steps add up to BIG changes.

I DO math.

- My questions are in process of being answered. I continue to put them out there and answers come. It is O.K. to ask several different people several different times. Each day I progress toward my goals. As I walk into my math classroom, my mind opens to math. I take in positive suggestions only, and I make my math class feel safe for facing challenges.

- Working together with others helps. When I laugh, my mind opens to understanding, so I often look for humor in math. I see the processes more clearly every day.

I do MATH.

- My new understandings of math become stronger. Understanding happens gradually. The patterns are there and my eyes spot them more quickly each day. The more I practice, the more connections in my brain happen. I release the MEAN MATH BLUES! Solving math mysteries brings me joy.

I DO MATH.

Pushing Your Limits

1. Write the eight Interventions for Negative Thoughts in your journal. Choose one of your negative thoughts from your list in Chapter 4. Choose one of the interventions and experiment with it. After you have done the intervention, evaluate and record its usefulness.

2. What if you acted "as if . . ."
   - You could do math?
   - You had a right to do math at your own speed?
   - You were a successful student?
   - You were going to pass your math class?
   - You will receive the degree you want?

How would you act—talk, think, walk, work, see, and hear—differently? What behaviors would you incorporate into your life? Journal about these possibilities.

3. Read the Math Affirmations aloud three or four times a week—more often at first. Post them where you will see them as you work. Keep a copy in your math book for review. (Change them, if you wish, to make them more effective for you. However, make certain that you keep them simple, positive, and in the present tense.) You may feel silly reading the affirmations aloud at first. Remind yourself that this is an effective method to consciously influence your unconscious belief system about math. If feeling silly is the worst thing that can happen, what do you have to lose? The gains could be enormous.

4. For a bit of math history and some fascinating discoveries, try Mastering Math's Mysteries, Chapter 6. You might be surprised, and you will get more practice with patterns. Act "as if" you are a successful math student looking at these exercises. What's the worst thing that could happen if you tried them?

Fibonacci Numbers

Mathematicians, both professional and amateur, are enthralled with "certain numbers" and "certain number patterns." Especially when those "certain numbers" show up in the real world time and time again, mathematicians revel in exploring and discovering all their possibilities.

The number π, pi, is one of those numbers. You may already know that π is related to circles. Another special number is named "e" and is related to growth and decay. We will not discuss either π or e in this book, although they are worthy of your time and fascination and have had entire books written about them.

We are going to look at a sequence of numbers in this chapter and an arrangement of numbers in the Appendix that have been popular over the years. The sequence is called the Fibonacci sequence and the arrangement of numbers is called the Pascal Triangle. The Appendix also shows how the Fibonacci sequence of numbers appears in the Pascal Triangle.

THE FIBONACCI SEQUENCE. In the year 1202, Leonardo di Pisa, a.k.a. Fibonacci for "Son of the Bonaccis," published a mathematical problem about rabbit breeding. The solution to his problem was a sequence of numbers, now called the Fibonacci Sequence. The sequence is:

   1, 1, 2, 3, 5, 8, 13, 21, . . .

Study this sequence and see if you can figure out the pattern. (Hint: You have to know the first two entries to get started. From there, compare each entry to the two preceding it. For example, compare the entry 2 to the entries 1 and 1 that precede it. Compare the entry 3 to the entries 1 and 2 that precede it. Do you see a pattern of adding two consecutive entries to get the next one? 1+1 = 2, 1+2 = 3, 2+3 = 5, . . .)

1. Using the pattern from above, write the next four terms of the Fibonacci Sequence in the blanks:

   <u> 1 </u>, <u> 1 </u>, <u> 2 </u>, <u> 3 </u>, <u> 5 </u>, <u> 8 </u>, <u> 13 </u>, <u>   </u>, <u>   </u>, <u>   </u>, <u>   </u> · · ·

Over the years since 1202, the Fibonacci and related sequences have fascinated mathematicians because they occur naturally in seashells, pine cones, pineapples, sunflower seeds, flower petals, and more. Manmade beauties such as Greek architecture, the Great Pyramid at Gizeh, the pentagram (five-pointed star), and countless paintings incorporate them too. In 1962, an association was formed to explore this phenomenon and to publish a journal called *The Fibonacci Quarterly.*

THE GOLDEN MEAN. A number called the Golden Mean, used often by artists and artisans, comes from a sequence of ratios made by dividing one entry of the Fibonacci sequence by the preceding entry, like this:

$$\frac{1}{1}, \frac{2}{1}, \frac{3}{2}, \frac{5}{3}, \frac{8}{5}, \frac{13}{8}, \frac{21}{13}, \frac{34}{21}, \ldots$$

2. Use your calculator to do the divisions in the preceding sequence. Do 1 divided by 1. Then 2 divided by 1. Then 3 divided by 2. Then 5 divided by 3, and so on. Write your answers as a sequence with commas between them. You will find your answers getting closer and closer to the number 1.618. . . . This number has been called the Sacred Ratio, Golden Ratio, and the Divine Proportion, as well as the Golden Mean.

Know Yourself

7

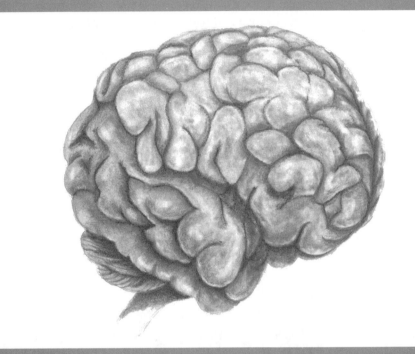

Let's Talk Smarts

RALPH WALDO EMERSON

"We are wiser than we know."

"Talent is habitual facility of execution."

There are *many* ways to be smart. Your intelligences and your abilities to learn are much more complicated and richer than you have ever dreamed. And you have an endless potential for incorporating new skills. Intelligence is not fixed—it can *change*. This chapter will help you understand how to use your unique intelligences to improve your math skills.

MULTIPLE INTELLIGENCES

An innovative and widely referenced model of human intelligence was developed by Harvard Professor Howard Gardner. Known as Multiple Intelligences, his work has changed the question "How *intelligent* are you?" to "*How* are you intelligent?" You are not smart in just one or two ways—but in many ways.

Gardner objected to assessing smarts by using I.Q. tests. Instead, he recommended looking at the ability to solve problems and produce meaningful products within a person's own culture. Gardner studied diverse cultures throughout the world to define intelligence in a manner useful for all types of human beings, whether they are artists, bush pilots, doctors, mothers, athletes, musicians, teachers, sailors, tribal chiefs, or engineers.

Gardner's first attempt to organize his massive research resulted in a preliminary list of seven intelligences. He later added an eighth intelligence—the Naturalist. He believes this list is not complete.

The figure on the next page shows the eight intelligences of Gardner's model. Copy the chart into your journal, leaving space for you to add notes that define each intelligence, assess your strengths, and recommend math study tips.

Facts about Multiple Intelligences

- All of your intelligences work together.

- Rarely does anyone exhibit only one of the intelligences.

- Verbal-Linguistic Intelligence and Logical-Mathematical Intelligence, commonly measured as I.Q., are *only two* of the many intelligences.

- There is now so much to learn that no one can learn it all, but there are necessary effective living skills that everyone should know, such as math.

- Last but not least: Intelligence can change!

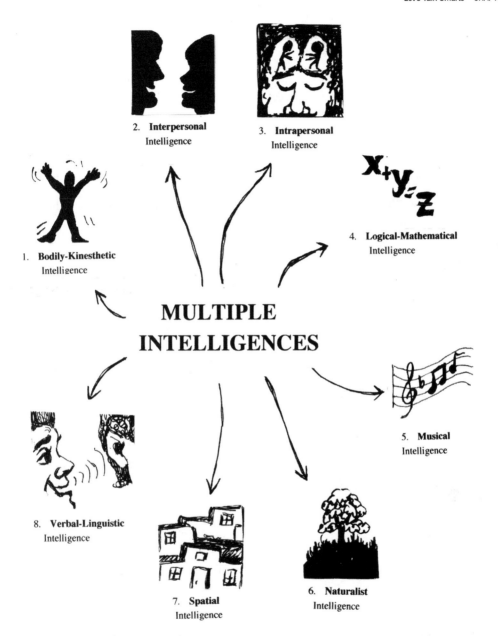

2.   **Interpersonal**
Intelligence

3.   **Intrapersonal**
Intelligence

4.   **Logical-Mathematical**
Intelligence

1.   **Bodily-Kinesthetic**
Intelligence

**MULTIPLE
INTELLIGENCES**

5.   **Musical**
Intelligence

8.   **Verbal-Linguistic**
Intelligence

7.   **Spatial**
Intelligence

6.   **Naturalist**
Intelligence

HOWARD GARDNER'S MULTIPLE INTELLIGENCES MODEL

Read these descriptions of each intelligence. As you read, consider how you would rate your skill in that area: strong, moderate, or weak. Shade the line from 1 to the number that you assess yourself for each intelligence.

1. BODILY-KINESTHETIC INTELLIGENCE. Ability to use the whole body or parts of it to develop products and solve problems as athletes, dancers, construction workers, actors, physical laborers, and surgeons do. (Famous examples: Jose Canseco, Roberto Clemente, Michael Douglas, Karim Abdul Jabbar, Michael Jordan, Mark McGwire, Shaquille O'Neal, Julia Roberts, Fernando Valenzuela, Tiger Woods)

1	5	10
Weak	Moderate	Strong

2. INTERPERSONAL INTELLIGENCE. Ability to understand motivations and inner workings of other people and to cooperatively lead or work with them. Teachers, mediators, negotiators, politicians, leaders, salespeople, and psychotherapists are examples. (Famous examples: Cesar Chavez, Henry Cisneros, Bill Gates, Mahatma Gandhi, Dolores Huerta, John F. Kennedy, Martin Luther King, Jr., Golda Meir, Gloria Molina, Eleanor Roosevelt, Mother Teresa)

1	5	10
Weak	Moderate	Strong

3. INTRAPERSONAL INTELLIGENCE. Ability in knowing and understanding one's own inner mental processes, effectively reflecting on thoughts, dreams, spiritual life, and motivations. Philosophers, authors, artists, psychotherapists, and many solitary individuals in all vocations exemplify this intelligence. (Famous examples: Teilhard de Chardin, Milton Ehrichson, Karen Horney, Carl Jung, Thomas Merton, Claude Monet)

1	5	10
Weak	Moderate	Strong

4. LOGICAL-MATHEMATICAL INTELLIGENCE. Ability to think mathematically and logically, as well as to analyze and reason scientifically. Accountants, inventors, repairmen, teachers, and engineers all put information together symbolically or practically using this type of intelligence. (Famous examples: Bhaskaracharya, Marie Curie, Thomas Edison, Albert Einstein, Lise Meitner, Isaac Newton, Emmy Noether, Ellen Swallow Richards, Julia Robinson, Chien-Shiung Wu)

1	5	10
Weak	Moderate	Strong

5. MUSICAL INTELLIGENCE. Ability in interpreting, performing, and composing music using melody, rhythm, and harmony. Composers, conductors, jazz musicians, music teachers, rappers, and cheerleaders exercise this ability as they create or use music in their work. (Famous examples: Desi Arnez, the Beatles, Leonard Bernstein, Charo, Miles Davis, Julio Iglesias, Wolfgang Amadeus Mozart, Tito Puente, Linda Ronstadt, Poncho Sanchez, Barbra Streisand, Ritchie Valens)

1	5	10
Weak	Moderate	Strong

6. NATURALIST INTELLIGENCE. Ability to see patterns and relationships in the natural world, classifying and discovering order. Scientists, biologists, botanists, and environmentalists exemplify this intelligence. (Famous examples: Rachel Carson, Jacques Cousteau, Charles Darwin, Albert Einstein, Rosalind Franklin, Barbara McClintock, Maria Mitchell)

1	5	10
Weak	Moderate	Strong

7. SPATIAL INTELLIGENCE. Ability to form an abstract model in their mind of the three-dimensional world and then solve problems using that model. People who do this well include astronauts, sailors, muralists, engineers, architects, surgeons, sculptors, and painters. (Famous examples: Judith Baca, Franklin Chang-Diaz, Leonardo da Vinci, Sam Maloof, Michelangelo, Antonia Novello, Ellen Ochoa, Auguste Rodin, Helen Rodriguez, Vincent Van Gogh, Frank Lloyd Wright)

1	5	10
Weak	Moderate	Strong

8. VERBAL-LINGUISTIC INTELLIGENCE. Ability with words in writing, storytelling, discussing, interpreting, and talking. Poets, writers, lawyers, talk show hosts, teachers, secretaries, and editors who form thoughts and use words skillfully in their work are examples of people strong in this intelligence. (Famous examples: Julia Alvarez, Maya Angelou, Agatha Christie, Emilio Estevez, Ernesto "Che" Guevara, Alex Haley, Langston Hughes, Edward Rivera, John Steinbeck, Victor Villasenor, Alice Walker, Oprah Winfrey)

1	5	10
Weak	Moderate	Strong

Summarize Your Assessment

Record how you assessed your multiple intelligences. Shade each row across to the number you rated yourself.

Bodily-Kinesthetic										
Interpersonal										
Intrapersonal										
Logical-Mathematical										
Musical										
Naturalist										
Spatial										
Verbal-Linguistic										

1	5	10
Weak	Moderate	Strong

Write down your top three intelligences.

1. _____

2. _____

3. _____

HOW TO USE YOUR SMARTS IN MATH

Read on for suggestions on how to use your strongest intelligences to improve your math skills. Remember: Intelligence can change.

- If you are strong in *Bodily-Kinesthetic Intelligence*, you will find movement useful in your math studies. Memory is triggered by location, so moving your body to different places as you learn new concepts helps you remember them. The gift of this intelligence will help you to lay out or act out problems physically using your muscles and movement. Even moving around randomly as you think over, discuss, and work on math problems helps. Working on a large chalk- or whiteboard will benefit you. Walking a large figure-eight shape (Sunbeck, 1996) (used by learning disability specialists) as you speak math aloud will cement learning, because this movement changes your state of mind, thus involving more functions in your brain.

- If your *Interpersonal Intelligence* is strong, you do well organizing study groups and facilitating others working together to discuss math problems. Taking a leadership role, you will feel more comfortable and find that you sponge up math skills in your interactions with others. With a group you enjoy, facilitate an open forum where questions are welcomed. You may wish to volunteer to tutor students with fewer math skills. As you do this, notice how your understanding increases as you explain ideas to others.

- Strong *Intrapersonal Intelligence* means you may enjoy contemplating math on your own at least part of the time. You might enjoy biographies of some of the famous mathematicians who were quite solitary and contemplative. Systems of thought and systems of social organization can be described in mathematical terms. You may even discover a way that human actions can be described by a mathematical system. Explore chaos theory—a branch of math that brings order to systems previously thought to be unordered.

- If you are strong in *Musical Intelligence*, you may find the mathematical descriptions of what happens in music very fascinating and revealing. Recognize how symbolic, mathematical, and logical music is. By learning music you have internalized a whole mathematical system that you use in your mind to create beautiful, new structures. Rhythm in music is an auditory manifestation of fractions. You intuitively and creatively put fractional parts together over and over to make whole measures. As you read or perform music, you create fractions of sound in time. It would be helpful for you to find another musician to tutor or mentor you in math or to work with as a study partner.

- Having strong *Naturalist Intelligence* means you notice patterns and relationships. Turn this point of view on math by noticing similarities, differences, and categories. Beginning with the whole and working down into the parts may benefit you. It will be extremely important for you to have a working knowledge of the "big picture." You may decide to invent new names meaningful to you for mathematical procedures or concepts. The biological or geological systems with which you are familiar are similar to mathematical systems. For example, the numbers on the real number line are a system with subcategories such as integers and rational numbers, the way plant and animal life are systems with subcategories such as family, genus, and species.

- If your *Spatial Intelligence* is strong, you learn well by making models of problems—models that you can manipulate and move in order to understand symbolic meaning. This intelligence is basic and essential to the study of math because many concepts relate to three-dimensional figures. Use clay or plastic straws to create geometric shapes used in math problems. Use model cars or airplanes to simulate the action in motion problems involving distances, rates, and times.

- If you are strong in *Verbal-Linguistic Intelligence*, you may be more comfortable reading math books that explain with more words and fewer symbols. You may wish to write math symbols in words to make sense of them and to see and appreciate their beauty. You will be happy to know that the use of words is essential to higher math, in which few symbols are used and ideas are often expressed solely in words. Higher mathematicians are wordsmiths too.

Pushing Your Limits

1. In your journal, list your eight Multiple Intelligences in order from your strongest to your weakest. What are your top three? How could you use these to improve the others? Be creative. Which of my suggestions can you use to improve your math smarts?

2. Write about your experiences with your I.Q. Have they been positive or negative? How can you think differently about your intelligences and your educational experience from the Multiple Intelligences perspective? How does this model free you and help you push your limits?

3. Use what you have learned about using your intelligences to write three short-term, achievable goals for yourself. Choose one of them to work on this week. Record your results in your journal.

4. Choose one of these three goals for this week. Evaluate and record your progress next week.

- Quiz yourself over last week's work.
- Ask questions when you are confused.
- Attend class daily.

5. Choose an action that will give you feedback to increase your math understanding. Next week evaluate the effectiveness of this action in your journal.

- Talk to a tutor.
- Work through examples in the textbook on paper and check them.
- Guess the next step the teacher will make during class.

6. Math students often dislike and avoid fractions. For some thoughts and suggestions for fractions from a Multiple Intelligences perspective, try Mastering Math's Mysteries, Chapter 7.

Introduction to Fractions—Adding & Subtracting

Motions, discussions, contemplations, songs, patterns, models, and words are the favored activities of the different intelligences. It is possible to do all of these activities with fractions. In the Mastering Math's Mysteries for Chapters 7 through 13, I will attempt to explain fractions through alternative methods. (Watch for the song!) Fractions are basic to math at all levels. Avoiding them only raises anxiety.

Accommodating your particular combination of intelligences will take creativity and determination on your part. But, if you persist, you can do it. You may need to dance, discuss, philosophize, sing, classify by pattern, make models, or verbalize information about fractions. The more you personalize the information, the more your mind will grasp the concepts. Then once you grasp them, find a way that works for you to practice enough to remember.

WHY FRACTIONS? Fractions exist because the real world is not a whole number. Daily we experience half sandwiches, halfway home, half dollars, and doing half of the homework. We know about a quarter of a football game, quarter of a dollar, and quarterly payments.

But when we write $\frac{1}{2}$ sandwich, $\frac{1}{2}$ of the way home, $\frac{1}{2}$ of a dollar (.50 of a dollar), $\frac{1}{2}$ of the homework, $\frac{1}{4}$ of a football game, $\frac{1}{4}$ of a dollar (.25 of a dollar), and $\frac{1}{4}$ payment, we panic and forget that we really know what that means.

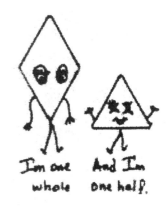

Im one whole

And Im one half.

WHAT IS *ONE HALF*? Every fraction has a "whole" related to it. The two couples to the right are correct examples of "one half" and "one whole." $\frac{1}{2}$ of a sandwich is one of two equal parts of a whole sandwich.

The **number on top** of a fraction (called the *numerator*) tells "how many" or the "number." The **number on the bottom** (called the *denominator*) describes what we have.

For example: $\frac{3}{8}$ or three eighths means that we have three of the pieces we get from splitting some whole thing into eight equal parts. $\frac{3}{4}$, or three quarters, means that we

Excuse me! Im one whole and he's one half.

have three of the pieces we get from splitting a whole football game, a whole dollar, or a whole year into four equal parts.

OPERATE ON THESE FRACTIONS. To add or subtract fractions, the denominators must be (and remain) the same because we need to add and subtract similar items. For example, $\frac{2}{4} + \frac{1}{4}$ is the same as saying two quarters of a game plus one more quarter of the game. That makes three quarters of the game.

Another example, $\frac{9}{10} - \frac{3}{10}$, is nine tenths minus three tenths, which is six tenths. Relating this problem to money, a dime is one tenth of a dollar. So $\frac{9}{10} - \frac{3}{10}$ becomes 9 dimes minus 3 dimes which is 6 dimes or $\frac{6}{10}$.

CHECK OUT THESE EXAMPLES. Here are more examples of adding and subtracting fractions:

(a) $\dfrac{4}{7} - \dfrac{1}{7} = \dfrac{3}{7}$ (b) $\dfrac{11}{25} + \dfrac{6}{25} = \dfrac{17}{25}$

(c) $\dfrac{7}{8} - \dfrac{4}{8} = \dfrac{3}{8}$ (d) $\dfrac{4}{x} + \dfrac{12}{x} = \dfrac{16}{x}$

Notice that each time I added or subtracted, the bottom (denominator) stayed the same and I did the computation with the tops (numerators), which were the number of items. Even when I did not have a clear description of the item in example (d), I still added the tops and kept the bottom the same. (The "x" was just a dummy stand-in for an unknown number. Since there was an "x" in both denominators, I could still add the tops and keep the bottom the same.)

The four examples I just gave you are very similar to:
(a) 4 dogs minus 1 dog is 3 dogs.
(b) 11 apples plus 6 apples is 17 apples.
(c) 7 books minus 4 books is 3 books.
(d) 4 cars plus 12 cars is 16 cars.

PRACTICE THIS KIND OF THINKING. Work these examples. If you feel squiggly, return to the parts of the explanation that helped you understand. Check your answers with the solutions in the Appendix.

1. $\dfrac{6}{10} - \dfrac{5}{10}$ 2. $\dfrac{17}{37} - \dfrac{12}{37}$ 3. $\dfrac{2}{8} + \dfrac{3}{8}$

4. $\dfrac{6}{12} + \dfrac{5}{12}$ 5. $\dfrac{13}{x} - \dfrac{3}{x}$ 6. $\dfrac{2}{12} + \dfrac{3}{12}$

8 Your Learning Mode

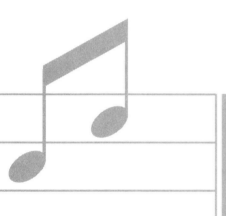

A successful life is the

unique invention of the

person who lives it.

PorchView
9/20/00

Quiz Yourself Before Reading On

Circle a, b, or c.

1. When you ask for directions, do you:
   (a) Remember them verbally in your mind?
   (b) Need a map or written instructions to follow?
   (c) Move your arms and point to review the directions before driving off?

2. When you are a student in a classroom:
   (a) Do you find it difficult to sit still and listen?
   (b) Do you listen carefully and find noise distracting?
   (c) Do you sit close where you can see what's going on and take notes?

3. When you are assembling new furniture, do you:
   (a) Move the pieces around and start putting them together immediately?
   (b) Read the instructions and look over the diagram?
   (c) Prefer to read the instructions aloud or have someone else read them to you?

4. Do you:
   (a) Spell rather well and see words and pictures in your mind?
   (b) Spell poorly but remember words from songs on the radio?
   (c) Prefer activities where you can move around and don't have to spell?

5. Would you be most likely to say:
   (a) "I see what you mean."
   (b) "I catch your drift."
   (c) "I hear what you're saying."

6. Would you most likely use the phrase:
   (a) "It slipped my mind."
   (b) "I don't recall."
   (c) "It appears I forgot."

7. For your birthday, would you most prefer:
   (a) Lots of cards?
   (b) Lots of phone messages?
   (c) Lots of high fives?

8. Would you rather:
   (a) Read a book?
   (b) Ride a horse?
   (c) Listen to the radio?

9. Would you rather:
   (a) Cook a meal?
   (b) Go to the symphony?
   (c) Watch a movie?

LEARNING MODES

Everyone accesses and processes information from the environment differently. In this chapter, you will attempt to discover more about how you do this. This discovery process is ongoing.

The Learning Modes model presented here is fairly well known and simplistic. Use this model to begin observing and experimenting with how you learn best.

A note of caution: Be careful of labels. They are static. You are not. If a label helps you become a "bigger" person with more alternatives, use it. When it does not benefit you anymore, shed it as a butterfly sheds a cocoon.

Often one of the three modes—Visual, Auditory, or Kinesthetic—is primary and is used more than the others. Some people believe that everyone is primarily either visual or auditory.

You are most likely a combination. You might have a primary and a secondary mode. The truth is that you use all three modes to make sense of the world around you.

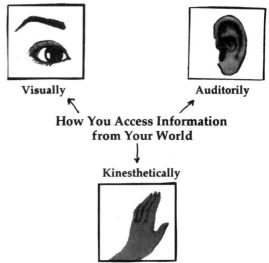

Visually   **Auditorily**

How You Access Information from Your World

Kinesthetically

YOU AND LEARNING MODES

Evaluate your quiz results from the beginning of this chapter by circling your answer to each question below and then adding the circles in each column.

	I	II	III
1.	b	a	c
2.	c	b	a
3.	b	c	a
4.	a	b	c
5.	a	c	b
6.	c	b	a
7.	a	b	c
8.	a	c	b
9.	c	b	a
Score:			

Shade up to your score in the corresponding columns on the next page. The highest column *may* be your primary learning mode and the second highest, your secondary learning mode.

	Visual	Auditory	Kinesthetic
9			
8			
7			
6			
5			
4			
3			
2			
1			
	Visual	Auditory	Kinesthetic
	I	II	III

Continue to evaluate whether this assessment fits you as you read the descriptions and examples of each learning mode in this chapter.

Read the sections on visual, auditory, and kinesthetic learners. Notice which mode has more activities that you habitually use. Mark those activities with a check mark. Mark others that you find promising with a question mark. The more activities you involve in your learning from all three of the modes, the more easily you will learn.

Use your primary learning mode to compensate for the other two modes and to make your other modes stronger. Do not give away control over how material is presented to you. Adapt material given in a different style to your own style. Experiment with the other modes to improve them.

KNOW THE MODES

Visual Learning

Visual learners require seeing. You form pictures and often see words spelled out or problems worked or situations happening in your mind. You may have a private movie screen in your head. When you understand something, you might say, "I see," and you do "see it" in your mind. You often forget verbal requests unless you write lists and reminders and put them where you can see them. You may speak slowly because of the difficulty of putting into words what you see in your mind.

Example of a visual learner: Sitting in front of the classroom, I see what's happening and am not distracted by watching other students. I take copious class notes, making little sketches to jog my memory. As a strong visual learner, I find that writing speeches down, recording them, condensing my speech to note cards, and then listening to my speech as I look at my notes or drive makes giving talks easier. Of course, I take my visual cues as backup. As a pianist since age six, I was "stuck" reading music—never playing without it—until I took a jazz piano class where I learned to visualize chords and keys. Frequently I clear my mind for work by clearing my work space first to make it look neat.

HELPFUL HINTS FOR IMPROVING VISUAL INPUT

- Sit in the front of classes or meetings so you can see everything.
- Develop skill at note taking to change verbal input into visual input.
- Sketch course content. Even the crudest sketch can help you remember ideas.
- List your tasks—even the ones you've completed, to have the satisfaction of visually crossing them out. (My husband laughs when I do this, but it works for me.)
- Use notes on your favorite colored stickies to help you remember.
- Close your eyes when you want to block out unpleasantness. (A fanatic about punctuality, I sometimes change the car clock when I'm late and then relax because it "looks like" I'm on time.)
- Evaluate the appearance of your study environment. Make it look conducive to learning. A well-placed poster or a desk turned away from clutter may work wonders in clearing your mind to study better.
- Write yourself encouraging messages and post them where you can see them.
- Picture yourself in situations where you have succeeded in the past.

Auditory Learning

Auditory learners rely on hearing. You listen to messages in your mind. You can repeat conversations or verbal input word for word. You often know all the words to songs you've heard. Radios, cassette players, and portable headsets play an important role in your life. The spoken word is essential. You may say, "I hear you" or "Sounds good" when you understand.

Example of an auditory learner: My artist friend, Emily, learns as she listens to tapes while painting and working. She is always conscious of her auditory background and travels with a cassette player and headphones. She tapes her textbook notes onto a cassette with her favorite music in the background. She often trades her artwork for someone else recording readings she wants to hear. Emily takes her tape recorder to the classroom and carefully chooses her location to avoid distractions from listening well. She can sing many '60s and '70s standards and has been known to recite 10-page poems "by ear."

HELPFUL HINTS FOR IMPROVING AUDITORY INPUT

- Choose the best classroom location for listening.
- Tape record the class session and listen to your tape.
- Ask questions in class and listen carefully to the replies.
- Read the textbook aloud to yourself as you study.
- Record your textbook or your class notes.

- Read your class notes aloud.
- Teach yourself to read aloud in your mind without making sounds. During exams, you can hear the test questions as well as see them. (If you need to move your lips, warn your instructor so you aren't accused of cheating.)
- Study with others. Talk about the course material.
- Tell others (or your pets) what you are learning in class. Mentally replay these conversations during exams.
- When you study, choose the auditory input in the background carefully. You are influenced by the sounds around you—especially talking. You may discover that you have favorite background sounds that help you concentrate.
- Use headphones so that the auditory input is of your own choosing.
- Consider using earplugs during exams to mask distracting noises. Inexpensive earplugs are available at drugstores. Notify your instructor in case she makes announcements.
- Speak positively to yourself during your work. Note the negative statements and identify the Thought Distortions.

Kinesthetic Learning

Kinesthetic learners need to move around and work manually with ideas. You touch things a lot. Smells and textures are important. You sometimes have difficulty sitting still in class just listening. The more activity you experience while doing a skill, the better you learn it. The more skin and muscles you use, the better you remember. Since most classrooms are taught by auditory or visual learners, you need to be creative to learn. Volunteer to move about the classroom acting as an assistant. Even small motions that seem unrelated to the activity such as swinging a leg, drawing, or knitting ease your integration of ideas.

Example of a kinesthetic learner: My social worker/teacher friend, Karin, moves through life touching and hugging. Interacting personally with her students, she keeps her adult classrooms active by using toys to illustrate ideas. As a student, Karin chose her teachers and mentors carefully so that her kinesthetic style was enhanced and supported. Her school projects involved activities to implement ideas. Cooking aromas, potpourris, and candle scents fill her comfortable home. The feel of her surroundings, not the look, is essential to her.

HELPFUL HINTS FOR IMPROVING KINESTHETIC INPUT

- Sit where you can actively participate in classroom events.
- Sit where you can move as needed without disturbing others.
- Draw pictures in class of the material being taught.
- Take notes creatively. Experiment with webbing (see Chapter 14).
- Ask and answer questions.
- Make models of the concepts whenever possible.

- Become skilled using your fingers and toes when doing math.
- Educate your instructor about kinesthetic learners and ask for assistance in developing models of the material with which you can interact physically.
- As you study, move around.
- Walk and talk to yourself about the material.
- Walk a figure-eight pattern, swinging your arms as you recite material you want to remember for your coursework. This walk will activate different parts of your brain and integrate concepts more fully.
- Work on the chalk- or whiteboard whenever you can.
- Incorporate pictures of models, if possible.
- Pat yourself on the back—physically!—whenever you do well.
- Make sure your pen and the writing materials that you use please you.
- Make physical comfort a priority as you study.

Whether you are a visual, auditory, or kinesthetic learner, always take charge of your own learning!

Pushing Your Limits

CHAPTER 8

1. Which learning mode do you use the most? Write in your journal three activities suggested for this mode that you might use to increase your learning. Pick three activities that you have not used before. Experiment with them.

2. In your journal, answer the question, "What does it mean to take charge of my own learning?"

3. Identify the activities on the worksheet included at the end of this chapter that will supply auditory, visual, and kinesthetic ways for learning fractions. How might you do something similar in the math that you are learning now? Write your ideas in your journal as they pop into your mind.

4. Mastering Math's Mysteries, Chapter 8, continues the subject of fractions. Besides using the activities that fit your strongest intelligences, see how you can utilize the three learning modes. Provide for visual, auditory, and kinesthetic input with these ideas. Review the fractions activities in Chapter 7 to see how your primary and secondary learning modes can work to give you a better understanding of fractions.

Working with Halves

Marie baked three pies for a family dinner. While she was out, her kids ate *half of a pie*. When she returned, she screamed, "Oh, no! Only _____ pies left! That is not enough." What did Marie scream? Think about it.

METHOD 1: IMAGINE OR DRAW OR BAKE WHOLE PIES.

- Marie had 3 pies.
- One half a pie was devoured by her children.
- That would be: 3 pies minus one half pie.
- They didn't touch 2 pies. Together they ate one half of the third pie, leaving one half.
- That would leave: $2 + \frac{1}{2}$ pies. Marie screamed: "Only $2\frac{1}{2}$ pies left!"

METHOD 2: IMAGINE HALF PIES. A second way to add or subtract one half is to recognize how many "half pies" there are in all of the whole pies and then add or subtract the one half.

Marie had three whole pies, which means she had six half pies, because each whole pie contains two half pies. When her kids ate one half pie, she was left with five half pies. Of course, she would most likely recognize those five half pies as two whole and one half pies.

Three pies = six halves or

$$3 = \frac{6}{2} \quad \text{so} \quad 3 - \frac{1}{2}$$

$$= \frac{6}{2} - \frac{1}{2}$$

$$= \text{six halves minus one half}$$

$$= \text{five halves}$$

$$= \text{two wholes and one half}$$

$$= 2\ 1/2 \text{ pies}$$

THINK DIVISION. Changing five halves into two wholes and one half can be accomplished by using division. Notice that it works to think of five half pies as five divided by two, which is two with one left over, or two and one half. Here are three other examples to contemplate or visualize:

$$\frac{9}{2} = 4\frac{1}{2} \qquad\qquad \frac{15}{2} = 7\frac{1}{2} \qquad\qquad \frac{24}{2} = 12$$

PICK YOUR FAVORITE. Which of the two methods—thinking of whole pies or of half pies—do you prefer? Practice adding and subtracting $\frac{1}{2}$. Sketch pie pictures to help you visualize. Talk yourself through to get these ideas into your ears. Make pies to get these problems into your skin and muscles. Think pies!

1. $4 - \frac{1}{2}$ 　　　　 2. $\frac{1}{2} + 4$ 　　　　 3. $3 - 1\frac{1}{2}$

4. $4 - 2\frac{1}{2}$ 　　　　 5. $4 + 3\frac{1}{2}$ 　　　　 6. $4\frac{1}{2} - 2$

Consider that we could generalize our work here to thirds, fourths, fifths, sixths . . . We could even say that Marie's kids ate $\frac{3}{100}$ of a pie. Not very likely, but it could happen. If they did, that means those kids cut one of Marie's scrumptious pies into 100 parts and ate three. That would leave two whole pies and 97 parts out of 100 in the third pie. So $3 - \frac{3}{100} = 2 + \frac{97}{100} = 2\frac{97}{100}$.

Imagine those sneaky kids and try these:

7. $4 - \frac{7}{8}$ 　　　　 8. $4 - \frac{2}{5}$

9

Who Can Do Math?

"What do you plan to do

with your one wild and

precious life?"

POET MARY OLIVER

"Don't give up until they kick you out."

EDUCATOR LINDA BROWNE

The title of this chapter is an interesting question. You may have answered it yourself many times and left yourself out. My answer to this question is, *"**Everyone** can do math."*

There are many different ways to appreciate and approach math. You will, no doubt, find yourself like most mathematicians. There will be parts of math that you will like and find easier than other parts. There will be strategies that you find useful and believe to be the best, and strategies that you dislike. You will notice that what you admit to liking about math is always just a little different from what others like.

The five approaches to math listed here are different ways of working with the subject.

1. Learning and creating theory
2. Applying mathematical concepts
3. Teaching mathematics
4. Playing with numbers
5. Using mathematics without calling it "math"

As we explore the five approaches to math below, you will meet six current and former Santa Ana College math student tutors. Even five years ago, if you had asked Enrique De Leon, Jazmin Hurtado, Sarah Kershaw, Joel Sheldon, Alex Solano, and Isabella Vescey if they had ever thought they would help teach math, they would all have replied, "No way!" When they started Santa Ana College, they took their math classes because they had to, not because they wanted to. Four of the six had to repeat courses.

Yet these six students not only successfully achieved their math goals, they became instrumental in the Santa Ana College math classrooms and Math Study Center. They became highly valued math tutors because of their ability to translate and explain mathematics. Santa Ana College students turned to them for feedback and were rewarded for their efforts. Touched by the magic of math, these six, in turn, touched others. Featured here, they share with you their experiences and insights in learning math. Look for their thoughts again, especially in Chapter 15.

Now let's look at the five approaches to math, in turn, and read about individuals who prefer them.

FIVE APPROACHES TO MATH AND PEOPLE WHO USE THEM

1. Learning and Creating Theory

Some people enjoy math theory. They appreciate the structure of math—the basic assumptions and definitions on which other mathematical relationships and ideas are logically built. You may have encountered some of this theory if you studied Euclidean geometry (regular high school geometry), which begins by defining a point, a line, and a plane. Theorists do not care if their math has practical applications. They just find the system beautiful.

- **Mathematician Julia Bowman Robinson (1919–1985)** was the first woman mathematician elected to the National Academy of Sciences and the first woman president of the American Mathematical Society. Fond of numbers since childhood (Reid, 1996), Julia was one of few math majors at San Diego State College in the 1930s. There, Julia read a book called *Men of Mathematics* and got her first taste of theoretical math. Excited by number theory, she pursued this interest at the University of California, Berkeley, in 1948, focusing on the tenth famous problem posed by mathematician David Hilbert. Over the years, Julia worked on Hilbert's problem, presenting papers on her conjectures and collaborating to explore one of her ideas, called the "Robinson hypothesis." Every year, before blowing out her birthday candles, Julia wished that someone would solve the problem during her lifetime—not necessarily her. Finally, in 1970, 22-year-old Yuri Matijascevic of Leningrad used the Fibonacci numbers (see Mastering Math's Mysteries, Chapter 6) to prove Julia Robinson's hypothesis and solve Hilbert's tenth problem. Julia was thrilled and wrote to Matijascevic saying, "I first made the conjecture when you were a baby and just had to wait for you to grow up!" (Albers, 1990).

- **Student/Math Tutor Sarah Kershaw** likes the creativity in math. She sees math as a construct of the human mind that becomes a language used to explain the world. She loves that there is more than one path to arrive at an answer, because she claims she is not very good at going the same way twice. Sarah pushes her limits every day, pursuing her bachelor's degree at University of California at Irvine with her focus set on a master's degree in divinity. Diagnosed with dyslexia as she worked her arithmetic problems in the second grade, Sarah did not learn to read until she was 12 years old. She first began to understand mathematics as she discussed theory with a calculus teacher. Recognizing that she learns best by working from the theory, or generalizations, down to the specifics, Sarah began devouring math books focused on theory and taught herself the math she needed to know. Today, in her 30s, she is a university honors student because she has discovered and mastered the school techniques that work for her. She has learned how to cope and approach learning when the symbols—whether letters or numbers—become squiggly as she examines them. At Santa Ana College, she skillfully tutors statistics and develops classroom demonstration tools to explain statistics concepts. She has lectured and trained math faculty and others at the college on working with students with learning disabilities.

- **Student/Math Tutor Enrique De Leon,** when asked what he likes about math, said, "I like going through a process in order to get an answer.

The process is actually more fun than the answer. I like the whole struggle of getting somewhere regardless of where it is, and I want to say I love math. It's fun. It's hard. It makes me pull out my hair—when I have some." When Enrique was my prealgebra and beginning algebra student, I frequently gave him my exam answer key after the test to proofread. He was exacting and took pride in detail. Working in groups at the chalkboard in prealgebra class, Enrique took the leadership, patiently explaining concepts to other students. Noticed and recruited as a Freshman Experience program math tutor, Enrique had not thought about teaching math. Now finished with three semesters of college calculus and sought after as a campus and private math tutor, he has a hard time remembering when he didn't want to become a math teacher.

- Math theorists that you may wish to research are **Lipman Bers, Olive C. Hazlett, Benoit Mandelbrot, Anna Pell Wheeler, Gottfried Wilhelm Leibniz,** and **Karl Friedrich Gauss.**

2. Applying Mathematical Concepts

Some people enjoy applying math ideas to the real world. They study or develop relationships called *formulas* that explain and predict natural phenomena, then use those formulas in building bridges, launching rockets, tracking trends, and all sorts of other important work. At more advanced levels, these people might be applied mathematicians, physicists, engineers, statisticians, architects, accountants, chemists, biologists, construction managers, or actuaries.

- **Chemist/Ecologist Ellen Swallow Richards (1842–1911)** earned, but was not awarded, a doctoral degree in chemistry from the Massachusetts Institute of Technology (MIT). Called the mother of ecology and a pioneer in environmental engineering, Richards used her chemistry expertise during her career to analyze water, gas, metals, minerals, air, and food to improve the environment of her community and eventually the world. While teaching at MIT for 38 years, Richards used her own home as a laboratory to experiment with ventilated heating systems, pure water systems, and waste disposal systems. She introduced the first sanitary engineering course at MIT and taught and mentored students who designed and operated the first modern sewage treatment plant. An authority on fires and a contributor to the development of one of the first gas ovens, she launched the sciences of consumer nutrition and environment education. A systems thinker ahead of her time, Richards' concern for home and community environments gained attention for serious public health issues.

- **Geneticist Barbara McClintock (1902–1992),** with ideas before her time, won a Nobel Prize for her revolutionary discovery in genetics. Born into a family that fostered her independence and creativity, McClintock earned a Ph.D. in botany by age 25 and did some teaching before becoming a longtime researcher at Cold Spring Harbor laboratory in Long Island, New York. It was there that she concentrated on the chromosomes of corn and developed her transposition theory, often referred to as the "jumping genes" theory. Reporting her findings to colleagues in 1951, she was greeted with total silence and rejection. A private person, she returned to her research. It was not until the 1970s that her theory was recognized for its significance. The transposition theory has been used to explain how genes mutate and has paved the way to discoveries in cancer research. Toward the end of her life, McClintock received the awards and accolades that finally recognized her detailed and painstaking research.

- **Student/Math Tutor Alex Solano** likes how math can be applied to the real world. He said, "What is amazing, and what I never thought about, is how much you can do with math. The higher up I have gone, the more opportunities I see with math. I really like trigonometry and how you can measure certain angles if you are given certain information." Planning to become a kinesiologist, Alex spent the day in shock when his math teacher asked him to be her teaching assistant. During his first year at Santa Ana College, Alex disliked math even though he went to class daily and did all of the homework. Intermediate algebra overwhelmed him, forcing him to repeat it. During the second time through, Alex discovered the patterns and his own love for math. He earned his first A's in math in intermediate algebra, trigonometry, and pre-calculus, and credits the strong algebra background he gained in intermediate algebra for helping him understand and pass calculus. The teachers that Alex has worked with trust and value Alex's assistance with math students.

- Others who apply math for you to research are **Albert Einstein, Richard Feynman, Shirley Jackson, Gertrude Elion,** and **Joyce Bell Burnell.**

3. Teaching Mathematics

Some people enjoy working with others and teaching them mathematical ideas and problem solving in schools from elementary through graduate levels. The challenge of truly gifted teachers is to look continually at math ideas from the multiple perspectives of their students to discover how students see those ideas and how to adapt teaching methods to reach more learners.

- **Astronomer/Professor Maria Mitchell (1818–1889)** was the first woman member of the American Academy of Science. She discovered and plotted a comet invisible to the naked eye in 1847, winning a medal from the king of Denmark. Following stints as a librarian and school teacher in her own school, Maria became professor of astronomy at Vassar (1865–1888) in charge of the observatory, where she continued her observations of the heavens, interpreting and recording what she saw and always searching for mathematical explanations.

An advocate of higher education for women, Maria helped found the Association for the Advancement of Women and mentored many future women scientists, including Ellen Swallow Richards. Maria's teaching methods included small classes, individual attention, hands-on experience, and plenty of mathematics.

- **Student/Math Tutor Joel Sheldon,** a bright, creative student who thinks outside the box, says he did a "whole lot of nothing" looking for a good time and working menial jobs in and after high school. Now in his mid-20s, he has found that "good time" by working in the math classroom as a teaching assistant and in the Math Study Center tutoring students from basic math through calculus. Joel remains on the fence about whether to major in mathematics or physics, but he is definite about his love of both. His easygoing, straightforward style with math students is popular with both instructors and the students he assists. Joel likes the fact that "math can be applied to everything that you see around you." He says, "Knowing that I can do it and that I can convey it to others is one of the things that helped me to like it more."

- **Student/Math Tutor Jazmin Hurtado** toys with becoming a math instructor eventually, even though her main goal is medical school. She sees her friends taking more math and finds that she is curious about what they are learning and misses the challenge. Jazmin likes math because it feels natural to her. She understands it and can picture it. She says, "Sometimes I get an answer but do not grasp the concept. That really bugs me, so I sit there and stare at the problem and do it over in my head for two or three hours until I get it. When I get it, I feel a very good satisfaction. I get that with math—not with any other subject. That's why I like math." Jazmin returned to school to follow her dream of becoming a pediatrician after a six-year break working as an office manager. Speaking only Spanish in Mexico from age 7 to 16, Jazmin maintained her English skills by daily watching the Disney channel on television. At Santa Ana College, Jazmin worked her way up from prealgebra, completing her second semester of calculus in the summer of 2000. Although the lower-level algebra and geometry courses were fairly easy for her, she discovered that she had to read the book and dig in more when she reached trigonometry. Forced to reevaluate and modify her work and school schedule, Jazmin found that she needed hours of study time—often more than her study partner did. Undeterred, she depended on persistence and honesty about whether she had really done what was needed to learn the material. A too-full schedule hindered her from passing second-semester calculus the first time through, but she stayed in until the end, taking detailed notes. Those notes enabled her to pass the next semester even though she had broken her foot just before the first exam and had not yet seen a doctor. A popular algebra assistant, Jazmin currently works in the college transfer center as she attends her University of California at Irvine classes.

- Consider a math concept that you know well, such as counting or adding. How would you teach that concept to someone else?

4. Playing with Numbers

Some people enjoy the patterns and relationships among numbers. Almost as a hobby, they love working puzzles or collecting little-known facts about numbers and number systems. Weaving fascinating tales, some of these people have written books with titles like *The Magic of Mathematics, The Man Who Counted, Powers of Ten, The Number Devil,* and even *Another Fine Math You've Got Me Into.* They delight in interesting little facts about certain numbers.

- **Author/Mathematics Teacher Calvin C. Clawson** explores the discovery process in math in his book *Mathematical Mysteries: The Beauty and Magic of Numbers.* In a chapter called "Exotic Connections," Calvin discusses the Fibonacci numbers, patterns, and radicals. (See Mastering Math's Mysteries.) In the introduction to his book, Calvin confesses that his "stomach takes a hop and [his] heart races" when he sees "numerous bizarre [math] symbols which [he's] not seen before." He says his reaction is the same as the "math phobic. The difference is that [Calvin] know[s] the feeling of discomfort will pass, that it is only an automatic response" to a new situation (Clawson, 1996, p. 3). Clawson has also written *The Mathematical Traveler: Exploring the Grand History of Numbers* and *Conquering Math Phobia.*

- **Student/Math Tutor Isabella Vescey** likes the feeling she gets when math flows for her—when she's comfortable doing it. There are times, she says, "when I do my homework and everything else is gone." She confesses that sometimes, while her family is watching television, she sits in the kitchen playing with the numbers and buttons on her calculator. A returning student and a 40-something mom, Isabella used to feel tearful and turn off during beginning algebra classes. In fact, she took the course four times before she passed. This year she shocked herself when her hand shot up after her statistics teacher asked who would be taking more statistics beyond the current course. Isabella recognizes in retrospect that she always could do math, but she did not realize that when she was younger. She now attributes her success to not being afraid to take risks on paper and to try new things. As a valued and candid math teaching assistant, Isabella has gained work experience that will help her obtain her vocational education teaching credential.

- People who love patterns will find that the number 142,857 is quite special (see the box on the following page). If you are a person who likes examining the special quirks of numbers, look at the *order* of the digits in the answers when 142,857 is multiplied by the numbers 1, 2, 3 . . . (Tahan, 1993).

5. Using Mathematics Without Calling It "Math"

Some people use math without thinking that it is math. Fractions—parts of a whole—are second nature to the chef, musician, artist, and athlete, who

The Number 142,857

```
 1 • 142,857 =   142,857
 2 • 142,857 =   285,714
 3 • 142,857 =   428,571
 4 • 142,857 =   571,428
 5 • 142,857 =   714,285
 6 • 142,857 =   857,142
 7 • 142,857 =   999,999
 8 • 142,857 = 1,142,856
 9 • 142,857 = 1,285,713
10 • 142,857 = 1,428,570
11 • 142,857 = 1,571,427
12 • 142,857 = 1,714,284
13 • 142,857 = 1,857,141
14 • 142,857 = 1,999,998
15 • 142,857 = 2,142,855        What's next?
```

Notice that when 142,857 is multiplied by the numbers from 1 through 6, the answer has the same digits in order from left to right, except some of them have gone "to the back of the line." Multiplying by 7 yields only 9s, but multiplying by the numbers 8 through 13 gives answers with the same digits as 142,857 in the same order; however, some have gone "to the back of the line" again *and* some numbers are written in parts. For example, 9 • 142,857 = 1,285,713, where the number 4 is replaced by a 3 and 1. Multiplying by 14 gives another result that is all 9s—except that one of them is written in parts (1 and 8). Interesting answers continue. Grab a calculator and keep multiplying to see what else happens. (If you don't see the patterns, come back later or show someone else. This is a good place to practice asking questions of another person.)

instinctively blend and creatively combine fractions. The chef knows the fractional amounts and relationships among ingredients that combine to form exquisite flavors of a "whole" dish. The musician knows the "whole" of a measure of music. The jazz soloist can play a combination of quarter, eighth, sixteenth, and thirty-second notes that magically add to one whole measure, bringing him to the down beat of the next measure at exactly the same moment as the rest of his group. The artist puts fractions together, visually creating three-dimensional delights on a "whole" canvas. Athletes add quarters to play a whole basketball or football game.

- **Artist/Musician Bob White** is a man of many talents. Besides having sung with the 60s musical group The Platters and still performing, Bob produces photographic-quality pencil drawings and teaches others to do the same. He assisted me with the pencil drawings that you see at the beginnings of Chapters 7, 9, 12, and 16. Bob has also worked as a draftsman and is a retired Los Angeles police officer. When I interviewed him about math, he told me to tell struggling math students, "Actually, if you can find a way in your life to get around having to use math, complicated math, do so. But I know that you can't get through life—you can't go through a single day—without math. It is involved in your everyday life."

- **Artist/Teacher/Studio Owner Phyllis Biel,** trained by seascape artist Earl Daniels, enthusiastically reports a lifelong love of painting. Operating an art studio for over 35 years where other artists gather daily to work with their students, Phyllis carves out time to produce exquisite, realistic oils. When I asked her advice for struggling math students, she bluntly answered, "Be an artist." In spite of her denial of her math skills and training, she knows that she completed algebra. As we concluded our interview, Phyllis pulled me to a cupboard containing her painting inspiration file and showed me a pile of artistic diagrams based on mathematical concepts—logarithms and exponents.

- **Public Administrator Don McIntyre,** former city manager of Pasadena, California, and former general manager of a large urban sewage treatment agency, said that it "never registered" with him "how much you are going to use [your math training] in your life's work no matter what you do." Don "found statistics very helpful to do analytical reports" and more as he guided the city of Pasadena through major redevelopment of Old Town Pasadena and Orange County Sanitation Districts through strategic planning and major reorganization of the agency. He said he would advise students to "encourage teachers to make [math] practical."

- Brainstorm about when you instinctively use fractions to create a "whole" product.

THE PLAYING FIELD

Opportunities in the science, engineering, and math fields have improved greatly for minorities and women in this country over the past century. Even so, there is still room for improvement for people from all backgrounds. Limitations on some because of race or gender are limitations on all.

One barrier for women because of their gender is speculation about women's intelligence in math. The area where this gender difference supposedly appears is spatial visualization. Many now believe that spatial visualization is a learned, not an inherent, ability. Playing with shapes and puzzles, studying mechanical drawing, making maps, studying art, playing certain computer games, and moving your body about in space as athletes do have been found to improve spatial visualization skills in both males and females.

The bottom line in intellectual differences is that the differences are as great within genders and races as they are between them. Any gender differences that have been identified have an effect only at very high levels of mathematics and do not affect those who are motivated enough to reach those levels.

Historically, other barriers were placed before women interested in the sciences. By 1993, only nine of the several hundred Nobel Prize winners in science were women. In her book *Nobel Prize Women in Science*, Sharon Bertsch McGrayne (1993) identified the following obstacles these winners faced because of their gender. Some women scientists were:

- Disrespected and called "unnatural beings."
- Barred from academic high schools or universities.
- Banned from science or math lectures and chemistry or physics labs.
- Accepted in research facilities only with a male partner.
- Passed over for paid positions by universities, but accepted as volunteers.
- Disallowed advanced degrees that they had earned.

McGrayne credits their families, churches, female educational institutions, love for science, value of education, and desire for learning for sustaining the nine women who pushed their limits and completed Nobel Prize–winning work.

I have found great inspiration in the stories of the people featured in this chapter and others like them who heroically push beyond expectations and social barriers. With both a historical perspective and conscious knowledge of barriers, it becomes our own choice whether to be influenced or not by the environment and the context we live in. Choice is powerful!

Though darkness seems warm comfort,
  hatchling knows the time
  to peck away an opening—
long labor—then she's free.

Or off from seed, once roots reach deep,
  one stem must find its way,
  its steady climb past stony obstacles—
long labor—though its destiny, it knows,
    is up and out, toward the sun.

And even pain can serve:
  in her own shell, what she cannot remove
  she builds a shield around:
long labor—but the product of her industry—
    a pearl.

by Victoria Stephenson,
English Professor,
Santa Ana College

Reprinted with permission.

Pushing Your Limits

CHAPTER 9

1. Which approach to mathematics fits most closely with yours? Had it occurred to you that there were different approaches? Explore your thoughts in your journal.

2. Which person featured in this chapter do you relate to the most? What are the similarities between what you know of that person and yourself? What are the differences? Even people we read about can mentor us.

3. Each of the people highlighted in this chapter is a hero because they have all pushed the limits of their surroundings. Any of us who go beyond the expectations and limitations of our cultures—family, school, work, or community cultures—is a hero. Write about how you are pushing your limits.

4. We only know about the heroes in this chapter because someone has recorded their lives. Look around you and notice heroic women and men who are pushing their limits. You may wish to record some of what they say and do, if only to share with your children and grandchildren. Heroism in all forms is inspiration for us to be at our best. We each have the potential to be a hero in our own lives.

5. Choose one of these three goals for this week. Evaluate your progress next week and record what works on your journal "Goals Page."

- Congratulate fellow students on their math successes.
- Review your class notes over last week's material.
- Begin your homework problems within three hours after class.

6. Choose an action that will give you feedback to increase your math understanding. Next week evaluate the effectiveness of this action. Record your results.

- Answer questions in class asked by the teacher and other students.
- Work through the original problem using the answer you get.
- Discuss problems with classmates.

7. Mastering Math's Mysteries, Chapter 9, returns to adding and subtracting fractions—a topic begun in Chapter 7. Continue to face fractions and push that barrier down.

More on Adding & Subtracting Fractions

How you understand fractions depends not only on your unique combination of intelligences and learning modes but also on your favorite approach to math, whether it is through theory, application, teaching, playing with numbers, or naiveté (i.e., doing fractions without calling them fractions). The theorists and the number fans most likely will understand fractions the way they are traditionally presented in school. The teachers will get caught up in the "how to" and the methodology. The practical people who like applications will want real-life examples such as those we have used in Chapters 7 and 8. Those who do fractions without calling them fractions will want to recognize the understandings that they already have, stay grounded in those understandings, and practice transferring their real-life knowledge to the numbers.

In this Mastering Math's Mysteries, we will discuss adding and subtracting when the denominators are not the same. This discussion is just a beginning for this topic, but it will, I hope, give you a foundation for continued work in a regular math textbook.

IT'S ALL IN THE GAME. Any sports fan knows that adding the first half of a game and one more quarter brings the game to the end of the third quarter. In mathematical symbols, that means $\frac{1}{2} + \frac{1}{4} = \frac{3}{4}$. There is really no way to use the numbers given in $\frac{1}{2} + \frac{1}{4}$ to get $\frac{3}{4}$ **except** to recognize that the first half is actually two quarters or $\frac{1}{2} = \frac{2}{4}$. Then a middle step allows us to add like this: $\frac{1}{2} + \frac{1}{4} = \frac{2}{4} + \frac{1}{4} = \frac{3}{4}$. When the bottoms or denominators become the same, we use the method for adding from Chapter 7. (Recall that method was to add the tops and keep the bottom the same.)

HERE IS THE BOTTOM LINE. We must get the same bottom (denominator) on fractions in order to add or subtract them. And it is simplest to use the smallest numbers possible, so we call the "best bottom number" the "Least Common Denominator" or LCD for short.

In our sports example, the 4 was the LCD for 2 and 4. To absorb this, readers who are kinesthetic, visual, bodily-kinesthetic, or spatial people need to draw a picture or make a 3D model using clay. Those students might need a pumpkin pie or pizza split into four equal parts to see that together two of those four parts can either be named two fourths or one half. This is the same idea that two quarters of a football game is one half. An excellent model for adding and subtracting fractions can even be made from an empty egg carton.

HOW TO TURN AN EGG CARTON INTO A FRACTION CALCULATOR. An egg carton from a dozen eggs is an easy model to make for fractions. The egg carton has 12 "pockets" for eggs—12 pockets for 1 *whole* dozen. Each pocket is one twelfth or $\frac{1}{12}$ of the *whole* carton.

Looking at your egg carton, ask yourself these questions:

- How many pockets would be in one third of the carton? Splitting the carton into three equal parts puts four pockets in each part. So four pockets or four twelfths would be one third, and $\frac{4}{12} = \frac{1}{3}$.

- How many pockets would be in two thirds of the carton? The answer is eight pockets or eight twelfths, so $\frac{8}{12} = \frac{2}{3}$.

- How many pockets would be in one fourth of the carton? Splitting the 12 pockets into four equal parts puts three pockets in each quarter. So three pockets or three twelfths makes one fourth. Or $\frac{3}{12} = \frac{1}{4}$.

- How many pockets would be in three fourths? Three fourths = nine pockets or nine twelfths.

If you have extra egg cartons, try cutting them into halves, thirds, fourths, and sixths.

1. How many pockets would be in one sixth? _____ Two sixths? _____ Three sixths? _____ Four sixths? _____ Five sixths? _____

PUT YOUR EGG CARTON TO WORK. Use the Egg Carton Calculator to add and subtract.

(a) $\dfrac{1}{3} + \dfrac{1}{12} = \dfrac{4}{12} + \dfrac{1}{12} = \dfrac{5}{12}$

One third of the carton plus one twelfth of the carton is four pockets plus one pocket, which is five pockets or five twelfths.

(b) $\dfrac{2}{3} - \dfrac{1}{4} = \dfrac{8}{12} - \dfrac{3}{12} = \dfrac{5}{12}$

Two thirds of the carton minus one fourth of the carton is eight pockets minus three pockets. The answer is five pockets or five twelfths.

Notice that *every fraction had to be renamed with a common bottom number (denominator) before adding or subtracting.*

PRACTICE MAKES PERFECT. Use your Egg Carton Calculator again.

2. $\dfrac{1}{3} + \dfrac{1}{4}$

3. $\dfrac{3}{4} + \dfrac{1}{12}$

4. $\dfrac{1}{6} + \dfrac{1}{3}$

5. $\dfrac{1}{6} + \dfrac{1}{4}$

6. $\dfrac{2}{3} - \dfrac{1}{12}$

7. $\dfrac{3}{4} - \dfrac{1}{3}$

8. $\dfrac{5}{6} - \dfrac{1}{4}$

9. $\dfrac{1}{3} - \dfrac{1}{4}$

CREATE YOUR OWN. Make up problems and check them with your Egg Carton Calculator. (Use only twelfths, sixths, fourths, thirds, and halves in your problems. The egg carton does have its limits.) Make up problems to add or subtract three fractions; for example:

10. $\dfrac{1}{6} + \dfrac{1}{3} + \dfrac{1}{4}$

PART FOUR

MANAGING THE MEAN MATH BLUES

Do the Math

10

Getting "In the Zone" with Math

"The quality of life

is much improved

if we learn to love

what we have to do."

**MIHALY
CSIKSZENTMIHALYI**

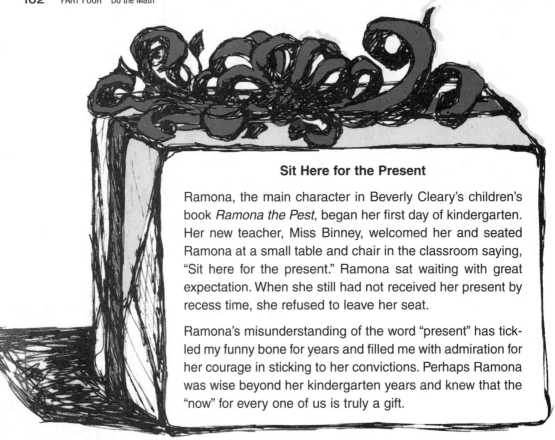

Sit Here for the Present

Ramona, the main character in Beverly Cleary's children's book *Ramona the Pest,* began her first day of kindergarten. Her new teacher, Miss Binney, welcomed her and seated Ramona at a small table and chair in the classroom saying, "Sit here for the present." Ramona sat waiting with great expectation. When she still had not received her present by recess time, she refused to leave her seat.

Ramona's misunderstanding of the word "present" has tickled my funny bone for years and filled me with admiration for her courage in sticking to her convictions. Perhaps Ramona was wise beyond her kindergarten years and knew that the "now" for every one of us is truly a gift.

Mental health professionals, philosophers, and, more recently, educators and scientists recognize that focusing on the *present moment*—right now—is the best way to deal with the past and with the future. Even one of artist/entrepreneur Mary Engelbreit's cards says: "Worrying does not empty *tomorrow* of its trouble. It empties *today* of its strength." This chapter will teach you how to focus on the *present math moment* and create more satisfaction and success with your math work.

FLOW

The word "flow" is used by psychologist and educator Mihaly Csikszentmihalyi (me-high chick-sent'-me-high) to refer to those moments that are satisfying, intensely focused, and timeless. Those moments include effortless concentration and detailed interest.

Athletes refer to this type of experience as being "in the zone"; **musicians** experience "flow" as they perform exquisite pieces of music; **artists,** as they create

To Create an Experience of "Flow" or "In the Zone"

1. Match your math skills with the demands of the math work.

2. Set clear intentions or goals for all math activity.

3. Get relevant and immediate feedback on how you are doing the problems and understanding the concepts.

in their studios. Most of us experience "flow" when we do something we love and are very good at.

Math tutor Isabella Vescey experiences "flow" and says she likes math when "math is working for me— that's when I'm comfortable doing it. It's flowing along nicely. I forget about everything else in my life and I'm just right there in the math. I like the feel." She also says that now in math, unlike before, "My mind doesn't wander. I show up for class every day and on time. I'm able to pay attention. In statistics, I noticed that I was paying attention rather than wandering off. That kind of amazed me."

Professor Csikszentmihalyi found that "almost any activity can produce flow provided the right elements are present." In fact, he found that adults experience "flow" more often at work than when they are doing leisure activities and that teenagers who study a lot experience more "flow" moments than teenagers passively watching TV.

Would you like your math work to be: Satisfying? Intensely focused? Effortlessly concentrated? Full of detailed interest? Timeless? Read on to learn how you can get in "flow" or "the zone" with math.

Match Your Skills to Your Work

The first element to create a flow experience is to **match your math skills with the demands of your math work.** Make certain that your math abilities balance what you are asked to do in your math course. Mathematics courses are skills courses where you are expected to perform at a specific math skill level coming into the course. Each math course depends on its prerequisite course.

If you have not studied the prerequisite course recently or did poorly in the prerequisite course, your skills will not meet the demands of the work. It is no secret to math teachers that students who passed the prerequisite course with a "C" grade seldom pass their current math class unless they make a major effort to shore up their weaknesses with tutoring and intense practice.

You will save time by taking all of the math courses needed for your major in sequence, one semester after another, because you will maintain your skill level and not have to repeat. Math students will be overwhelmed in math classes that are beyond their current skill level. Their growth level will be at a more elementary level than what they are asked to do. They will be missing essential skills for the current work.

Enrolling in a math class beyond your skill level and not doing the necessary work to raise your skills is guaranteed to increase your anxiety level. The key to speeding up in math is to slow down and understand the concepts as you move forward.

Students who are frequently late for or absent from class will find that their math skills do not meet the demands of the math work, because they have missed part of the development of new skills. Even missing 10 minutes at the beginning of a class period can hinder understanding.

Set Clear Goals

The second element to creating flow experiences is to **set clear goals and intentions for all math activities.** Professor Csikszentmihalyi says, "[You] can set goals for even the most despised task."

Goals such as high grades or completing degree requirements are too far from your day-to-day experiences in your math class. If they are your *only* two goals, they take away from the concentration necessary for working math. Each time you don't understand a concept or become confused, you are at odds with your future goals. **Set more *immediate or short-term goals* along with long-range goals to bring your attention and intention to the math process going on right now.**

The goals listed in the box are achievable and measurable daily. They support the broader goals of receiving a high grade and completing the course. They shift your attention and intention to *accomplishment now.*

I have suggested these goals to you in "Pushing Your Limits" sections of preceding chapters. If you have not already done so, make a journal "Goals Page" and write down the goals that you set. Compare those goals to the ones listed in the following box. Which goals were the most effective for you? Which goals are you willing to set and follow through today?

Now that you know the importance of clear and immediate goals, set three or four each week of your math class. Note and record your progress at the end of each day to bring satisfaction with math into the present.

Short-Term Goals

These goals get you "in the zone" with math. Check those that you might consider.

❏ Work three review problems each day to boost your math confidence.

❏ Recognize how much more math you know now than you did two weeks ago.

❏ Note your thought distortions and rephrase negative thoughts.

❏ Cheer yourself and fellow students as you learn new ideas. Smile in math class.

❏ Relax consciously in math class by breathing deeply.

❏ Summarize what you learned in class today.

❏ Write five questions on what you don't understand in this chapter.

❏ Review your class notes for last week's material.

❏ Quiz yourself on last week's work.

❏ Write down two problems each day that could be on the next math exam.

❏ Attend class daily on time.

❏ Take notes so you know what the teacher considers important.

❏ Copy everything the teacher writes on the board into your notes.

❏ Practice patience with yourself when new processes are presented in class.

❏ Mark where you don't understand your notes and textbook and ask questions.

❏ Complete 90% of the assigned homework with understanding.

❏ Plan time for getting assistance, and then do it.

❏ Speak personally to the teacher to establish rapport.

❏ Introduce yourself to three classmates to develop a math support system.

❏ Start or attend a study group.

❏ Bring problems to class with questions. If they don't get answered, find another way to get answers.

❏ Turn in completed homework on time.

❏ Begin homework problems within three hours after class.

❏ Study in the math tutoring center.

Get Feedback

The third element of creating a flow experience is to **get relevant and immediate feedback on how you are doing problems and understanding concepts.** Because of great individual differences, you will, no doubt, see math problems differently from how others, including teachers, see them. For that reason, to understand math *completely*, continuous "feedback" is essential.

Feedback **is clear and timely input on whether you understand correctly or not.** It shapes your growing understanding as you learn. Feedback is information on how well you understand the ideas, concepts, and problems in your math work. Without feedback you will not know whether your understanding is correct or not.

I have been suggesting these actions in "Pushing Your Limits" sections in preceding chapters. Write down those actions that you chose to take and eval-

Feedback Activities

These actions provide feedback about whether or not you understand. Check those you might consider.

❏ Teach someone else how to do your homework problems.
❏ Summarize the procedures and main ideas of a problem.
❏ Solve a problem several ways.
❏ Ask yourself if the answer makes sense.
❏ Work through the original problem using your answer.
❏ Check your answers against the answers in the back of the textbook.
❏ Work the examples from class over until you can do them without consulting your notes.
❏ Share your work with the class.
❏ Show your work to someone who knows how to work the problem.
❏ Talk over a problem with a tutor.
❏ Visit the teacher during office hours.
❏ Consult your study group.
❏ Discuss problems with classmates.
❏ Ask questions.
❏ Tutor other students.
❏ Answer questions in class asked by the teacher and other students.
❏ Copy the examples the teacher gives in class, then work them on your own.
❏ Draw a picture of the problem situation.
❏ Chart or graph the information in a problem.
❏ Work through examples in the textbook on paper and check them.
❏ Create a physical model of a concept or problem.
❏ Guess the next step the teacher will do during class.
❏ Solve a problem with a group at the board.

uate their effectiveness. Which ones worked the best for you? Which ones did not work well? Which actions are you prepared to take now that you recognize their importance? How has getting relevant and immediate feedback increased your understanding of math?

Note: Lots of feedback is necessary to logically order your learning in your mind. Only you, yourself, can be certain you get enough.

GETTING "IN THE ZONE" CREATES A POSITIVE MATH CYCLE

You get "in flow" or "in the zone" with math by (Csikszentmihalyi, 1997):

1. Making sure your math skills are up to the demands of your math class,
2. Setting immediate goals for your math activities, and
3. Getting all the feedback you need to understand the problems and concepts, you bring focused attention to each math moment. This can create a positive math cycle, as shown in the illustration.

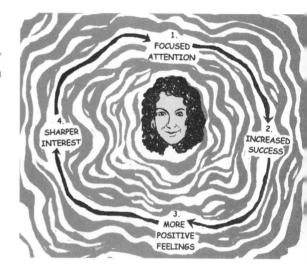

And around the cycle you go!

Pushing Your Limits

1. Define "flow." Write down times in your life when you experienced flow. How did you match your skills with the demands, achieve goals, and get feedback?

2. Honestly, have you placed yourself in the right level of math so that your skills match the course demands? Your college's placement test, a math teacher, or a counselor can help.

3. In your journal, write four immediate math goals you can achieve this week. Set a time when you will evaluate whether or not you achieved your goals. Which goals that you have tried so far have been most useful? Review the "Goals Page" in your journal.

4. Read the "Actions for Feedback" pages in your journal, where you have been recording and evaluating suggested actions. Think about how you get feedback or input on how you're doing in math. In your journal, write six more ways that you can increase the feedback you're getting.

5. To "Get in the zone" with fractions, work on Mastering Math's Mysteries, Chapter 10. As you work, set a few immediate goals that you can achieve right away, such as read the material aloud, work the examples with pencil and paper, use an egg carton to make models of the problems, or ask someone else for assistance. Determine ahead of time how you will get feedback or verification that you correctly understand.

Multiplying Fractions

The "why it works" of multiplying or dividing fractions is more difficult to explain than the "how to do it." I will attempt to explain multiplication here and division in Chapter 11.

REVIEW. Using the egg carton calculator that was introduced in Chapter 9, recall that each pocket was "one twelfth" or $\frac{1}{12}$ of the whole carton. Thinking about the 12 pockets, remember how other fractions could be represented like this:

$\frac{1}{6}$ of the carton was two pockets or $\frac{2}{12}$.

$\frac{1}{4}$ of the carton was three pockets or $\frac{3}{12}$.

$\frac{1}{3}$ of the carton was four pockets or $\frac{4}{12}$.

$\frac{1}{2}$ of the carton was six pockets or $\frac{6}{12}$.

MULTIPLY THOSE FRACTIONS. I will explain a few examples of multiplying using the egg carton, which accommodates twelfths, sixths, fourths, thirds, and halves quite well. (That is because all of the numbers 12, 6, 4, 3, and 2 divide evenly into 12 or one dozen.) With these examples, I hope that you will understand *why* we get the answers that we do. **Hands-on work with egg cartons is exactly what many minds need to understand. The model and motion**

ground the ideas in real life. I hope you do what you need to do to work through and absorb these examples.

(a) $\frac{1}{3} \cdot \frac{1}{2}$ This problem can be read "one third times one half" or "one third of one half." Look at one half of the egg carton. That "one half" contains six pockets. If we split that "one half" into three equal parts, two pockets would be in each third. One of those thirds would be two pockets, or two twelfths, or one sixth of the whole carton. Written out, we have calculated: $\frac{1}{3} \cdot \frac{1}{2}$ is $\frac{1}{3}$ of $\frac{1}{2}$, which becomes $\frac{2}{12}$ or $\frac{1}{6}$.

Conclusion: $\frac{1}{3} \cdot \frac{1}{2} = \frac{1}{6}$

(b) $\frac{2}{3} \cdot \frac{1}{2}$ This problem is similar to problem (a). Split one half of the carton into three equal pieces. Two of the thirds put together contain four pockets. Those four pockets are four twelfths or one third of the whole carton. So $\frac{2}{3} \cdot \frac{1}{2}$ is two thirds of one half, which is four pockets, or four twelfths, or two sixths, or one third of the whole carton.

Conclusion: $\frac{2}{3} \cdot \frac{1}{2} = \frac{4}{12} = \frac{2}{6} = \frac{1}{3}$

(c) $\frac{1}{6} \cdot \frac{1}{2}$ This problem asks for one sixth of one half. Splitting one half of the egg carton into six equal pieces puts one pocket in each sixth. So one sixth of one half is one twelfth of the whole carton.

Conclusion: $\frac{1}{6} \cdot \frac{1}{2} = \frac{1}{12}$

(d) $\frac{1}{4} \cdot \frac{2}{3}$ This problem asks for one fourth of two thirds. From previous work, we know that 2/3 of the egg carton is eight pockets. Splitting those eight pockets into four equal pieces, notice that each piece has two pockets. This means that one fourth of two thirds is two twelfths, or one sixth of the whole carton.

Conclusion: $\frac{1}{4} \cdot \frac{2}{3} = \frac{2}{12} = \frac{1}{6}$

SUMMARIZE WITH A TWO-FINGER MOTION. Notice that you could do each of the multiplication problems that we worked without the egg carton. The answer can be found by multiplying the tops of the fractions and also multiplying the bottoms of the fractions. Sometimes the answer could be simplified further, but an initial answer comes from multiplying right straight across—top times top and bottom times bottom. Kinesthetic learners or students with bodily-kinesthetic intelligence may want to hold two fingers of their *right* hand pointing straight ahead and move them to the right. This will remind them to multiply fraction "right straight across."

DISTINGUISH BETWEEN PROCEDURES. Doing the kinesthetic motion suggested previously while saying the words "Multiply fractions right straight across" can help students distinguish between the procedures for adding, subtracting, multiplying, and dividing. Only adding and subtracting procedures keep the bottom unchanged. The multiplication procedure multiplies straight across, changing the bottom. The division procedure changes the bottom too, in a different way—see Chapter 11.

Review the previous four examples to see that multiplying straight across does give the same answer.

(a) $\frac{1}{3} \cdot \frac{1}{2} = \frac{1}{6}$ because 1 • 1 is 1 and 3 • 2 is 6.

(b) $\frac{2}{3} \cdot \frac{1}{2} = \frac{2}{6}$ because 2 • 1 is 2 and 3 • 2 is 6. ($\frac{2}{6}$ is the same as $\frac{1}{3}$.)

(c) $\frac{1}{6} \cdot \frac{1}{2} = \frac{1}{12}$ because 1 • 1 is 1 and 6 • 2 is 12.

(d) $\frac{1}{4} \cdot \frac{2}{3} = \frac{2}{12}$ because 1 • 2 is 2 and 4 • 3 is 12. ($\frac{2}{12}$ is the same as $\frac{1}{6}$.)

TRY THESE EXAMPLES. Try them with the egg carton and by multiplying straight across. (Use the two-finger motion as you do them. Be sure to use your *right* hand.)

1. $\frac{1}{3} \cdot \frac{3}{12}$ 2. $\frac{1}{2} \cdot \frac{1}{3}$ 3. $\frac{1}{3} \cdot \frac{3}{4}$ 4. $\frac{1}{5} \cdot \frac{5}{6}$

5. $\frac{1}{9} \cdot \frac{3}{4}$ 6. $\frac{2}{3} \cdot \frac{3}{4}$ 7. $\frac{1}{8} \cdot \frac{2}{3}$ 8. $\frac{1}{2} \cdot \frac{5}{6}$

11

Questions and All That "Shy" Stuff

"I once had a garden

filled with flowers that

grew only on dark

thoughts, but they need

constant attention and

one day I decided I had

better things to do."

BRIAN ANDREAS

ASKING QUESTIONS

Why have a whole chapter about questions? The reason is that asking questions of:

> yourself
> your peers and colleagues
> your teachers
> your textbooks
> your tutors

asking questions, asking questions, asking questions is essential to learning—especially to learning mathematics. Asking questions honors your unique learning process, because no one else will know or have the same questions you have. Asking questions is a fundamental right in learning math and, frankly, there is no other way to learn. One small question can mean the difference between a "Eureka, I've got it" experience and an "I don't get it at all" experience. Which do you prefer?

Give Yourself Two Gifts:

1. Assert your right to ask questions.

2. Trust that the questions bubbling up in your mind are the right questions for right now.

Useful information results from questions starting with words like:

What if . . . ?	Where would . . . ?
What caused . . . ?	Tell me about . . . .
When would . . . ?	How do . . . ?
How could . . . ?	What's the difference . . . ?

Try These Powerful Questions

What if we tried this?	When would this process work?
What caused this step?	Tell me about this piece.
Where would this happen?	How do I recognize the difference?
How could this happen?	What else might work here?

The Tortoise and the Hares

A colleague told me a story about a woman in one of his graduate math courses who consistently asked questions that seemed very basic, almost too simple. My colleague and the other students began to roll their eyes and secretly smile whenever she raised her hand. They even quietly whispered to each other, wondering how she got into such a high-level math course. When final grades were revealed, there was only one outstanding score on the final exam and in the course. It was hers.

Questions pay off. They clarify your understanding. They help your brain wire the correct pathways in your mind—pathways you use over and over to learn deeper and more difficult material.

Who Has the Last Laugh?

In my mind's eye, I still see Michelle sitting in the front row of my geometry class 16 years ago, asking question after question. Sometimes students sitting behind her snickered and laughed because her questions were so intense and the answers seemed so obvious. She didn't even appear to notice the others, because she was so concentrated on her own learning. I admit that as Michelle's teacher, I had concerns about her future in math until three semesters later when I saw her in the hallway carrying a calculus book. She reported that she had successfully passed class after class of math requirements and was now completing her final semester.

Michelle taught me a great lesson with her focused style. As a teacher, I learned I could not predict success for my students based on their apparent ability. I learned that persistence counts more, and I wondered how far my snickering students with the higher grades had come.

ME AND SHY

When I was young, I called myself "shy." Believing I was shy became a reason for me not to talk to others and not to ask questions when I was in a new and strange place. Sometimes forcing myself to speak up, I would blush or sweat or get prickles all over my body or feel foolish. Sometimes I would tell myself how stupid or silly I looked or acted when I spoke. (What Thought Distortions were in my thinking?)

When I didn't speak up, I frequently went home with a huge headache—as if all those words I didn't say were stuffed inside my head pushing to get out. I was so conscious of myself—I wanted to say the RIGHT thing—but when I finally thought of the RIGHT words, the RIGHT time was past and the RIGHT words weren't RIGHT anymore.

The Boyfriend

As I Charleston-ed my way through the dance rehearsal for the high school musical, I noticed him watching me from the gym floor. The next day he knocked at my front door asking me to a movie. I was ecstatic! The brother of a classmate, he was older, tall, and cute. I was in awe of him—so in awe that I couldn't think of one single thing to say during our night out. For years I felt foolish every time that long, silent excruciating date flashed into my mind. One day I suddenly realized that *he* hadn't talked either. What a relief! The responsibility for conversation wasn't all mine. It was a shared responsibility and neither of us took it.

In my late twenties, I noticed that others in meetings and social groups often didn't say anything either. I began to experiment with talking to people. I noticed that when I talked, they talked too. Often they seemed to be waiting and were greatly relieved when I "broke the ice." I decided to drop the label "shy" for myself.

I noticed that the blushing and prickles and headaches and other physical reactions slowly disappeared as I practiced talking and asking questions. They were uncomfortable sensations, but never fatal.

I began to "act as if" I weren't shy, and eventually I wasn't. Amazingly enough, the things that I said seemed more RIGHT because I didn't pressure myself so much. I clearly followed conversations since I wasn't observing them from the outside.

There are still some situations where I *feel like* retreating to my passive past. I am quiet by nature and enjoy solitude. However, I recognize now that "being quiet" is a choice. Sometimes I choose to be quiet and other times I choose to participate. I make my own comfort level whether I am in a large room with hundreds of people or in a small room with one or two others. Conversing or not is always a choice.

YOU AND SHY

U.S. News and World Report magazine (Schrof, 1999) reported that one out of eight people experiences shyness to the degree that the person dreads encounters with other people. The symptoms—racing heart, dry mouth, sweaty palms, lack of words, feelings of confusion, and desire to escape—seem overwhelming.

If you are that one out of the eight, you don't need to give up hope. Consider these points.

LABELS LIMIT PEOPLE. Language is powerful. If you say you are shy, you equate *feeling shy* with *being shy*—a Thought Distortion from Chapter 5. You give yourself reasons to choose shy behaviors. Refusing to label yourself shy, you can imagine how people who are not shy act. Experimenting with new behaviors and "acting as if" you are not shy can foster permanent changes.

CHANGE IS CONSTANT. Brains and bodies are dynamic. They change constantly, and we have influence over those changes. Recall that your thoughts, behaviors, emotions, and body sensations are all interrelated. Your thoughts and your behaviors are the most powerful change agents you own. As you practice talking to people and asking questions, choosing safe and supportive places to do that, you will notice your comfort and confidence levels rising.

THESE THOUGHTS AREN'T TRUE. Negative self-talk like the following produces the body symptoms described above.

> Everyone in the room is always watching me.
> Everything I do or say is judged by others.
> I always make a fool of myself.
> I must never make a mistake.
> I must say or ask the right thing.

Notice that these five statements contain the Thought Distortions: Absolute Thinking, Overgeneralizing, and Shoulding. Notice how unrealistic these statements are. The words "everyone," "everything," and "always" are absolutes. How true can it be that "everyone" is "always" watching me and "everything" I do or say is judged and I "always" make a fool of myself? The word "must" makes two of these statements sound like orders and, frankly, they ask for behaviors that are perfect, not human.

NO QUESTION IS STUPID. Talking with your instructor and fellow students, you will want to watch your language. For example, when you precede questions with "This may be a stupid question, but . . . ," you dishonor your own learning process. You anticipate that others will think your question "stupid." That expectation could be fulfilled, but someone else dishonoring your question is his problem, not yours. When you dishonor your own thought process, that is *your* problem. Asking questions without this prefix gives you the respect you and your questions deserve.

WE AREN'T THAT IMPORTANT. Ah, yes. Believe it or not, the rest of the world has better things to do than watch us all the time. Notice that most people are more concerned with themselves than with you. They wonder how *you* see *them*. They busy themselves thinking what they will say next. Fortunately, or unfortunately, you just aren't so important to others that you become their focus of attention for more than a short time.

HELP IS OUT THERE. A therapist can help immensely if feeling shy is a severe and debilitating problem. Many colleges and universities have counseling assistance as part of their health centers. Life is too short not to participate in it fully.

QUESTIONS UNLOCK THE UNKNOWN

"I now ask small questions in my math class. At times I'll get the general concept of what's being taught, and, before, I would never ask those little questions that would help me to clarify the bigger picture."

—*Andrea, algebra student*

"We assume that what's familiar is simple when it really isn't. It took thousands of years to develop [math] and bring it all together. It didn't just pop into Newton's head one day. Each bit builds on the other. We assume because we are teaching it to people who are in grade school and high school that it must be easy. This represents at least 8,000 years of development."

—*Sarah Kershaw, student/math tutor*

Pushing Your Limits

CHAPTER 11

1. Make a list of statements you say to yourself that keep you from asking questions or speaking up in your math classroom. Think over each one of them and decide whether you want to keep this belief or not. Does this belief contain any Thought Distortions? What would an alternate belief be? Could you choose to make the alternate belief part of your thought process?

2. Identify and record four reasons for you to ask questions in your math class. Knowing these reasons can remind you how important questions are to your learning process.

3. Make a list of resources where you can get answers for your questions. Include people and locations by name so that you can refer to this list when you are frustrated. At stressful times when emotions flood your brain, written strategies can solve your dilemma.

4. Choose one of these three goals for this week. Evaluate and record your progress next week.

- Write two problems per day that might be on the next math exam.
- Note your thought distortions and rephrase your negative thoughts.
- Plan time for getting assistance and then do it.

5. Choose one of these actions that will give you feedback to increase your math understanding. Evaluate its effectiveness in one week in your journal.

- Check your answers with the answers in the back of the textbook.
- Show your work to someone who knows how to work the problem.
- Solve a problem with a group at the board.

6. Mastering Math's Mysteries, Chapter 11, continues work with fractions. Recall the value of asking questions from your reading in this chapter. Practice being courageous. Take all the time and ask all the questions that you need. "Be a Tortoise. Be a Winner."

Dividing Fractions

GET READY. To understand dividing fractions, first consider dividing whole numbers. "Six divided by two" can be thought of as "How many twos are there in six?" "One hundred divided by ten" can be thought of as "How many tens are there in one hundred?" We will think about fraction division in this way.

We will also use the Egg Carton Calculator introduced in Chapter 9. Remember that:

$\frac{1}{12}$ is one pockets.

$\frac{1}{6}$ is two pockets or $\frac{2}{12}$.

$\frac{1}{4}$ is three pockets or $\frac{3}{12}$.

$\frac{1}{3}$ is four pockets or $\frac{4}{12}$.

$\frac{1}{2}$ is six pockets or $\frac{6}{12}$.

DIVIDE THESE FRACTIONS.

(a) $\frac{1}{3} \div \frac{1}{12}$ Think of this problem as "How many one twelfths are there in one third?" Using the egg carton, recall that $\frac{1}{3}$ is four pockets or $\frac{4}{12}$. Also recall that $\frac{1}{12}$ is one pocket. Then rephrase the problem, "How many one twelfths, or pockets, are there in one third, or four pockets?" The answer is 4.

Conclusion: $\frac{1}{3}$ divided by $\frac{1}{12}$ is 4.

(b) $\frac{2}{3} \div \frac{1}{12}$ Think "How many one twelfths are there in two thirds?" Looking at the egg carton split into three equal pieces, notice that $\frac{2}{3}$ has eight pockets or eight twelfths. The problem "$\frac{2}{3}$ divided by $\frac{1}{12}$" becomes "How many one twelfths, or pockets, are there in two thirds, or eight pockets?" The answer is 8.

Conclusion: $\frac{2}{3}$ divided by $\frac{1}{12}$ is 8.

(c) $\frac{2}{3} \div \frac{1}{6}$ Think "How many one sixths are there in two thirds?" Recall that $\frac{2}{3}$ is eight pockets of the egg carton. Also recall that $\frac{1}{6}$ is two pockets. This division problem then becomes "How many twos are there in eight?" The answer is 4.

Conclusion: $\frac{2}{3}$ divided by $\frac{1}{6}$ is 4.

(d) $\frac{3}{4} \div \frac{3}{4}$ Think "How many three fourths are there in three fourths?" The answer has to be one.

Conclusion: $\frac{3}{4}$ divided by $\frac{3}{4}$ is 1.

SUMMARIZE. The preceding examples can be generalized into a procedure.

(a) $\frac{1}{3} \div \frac{1}{12}$ **is 4.** Notice that by "flipping over" the fraction on the right, $\frac{1}{12}$, to get $\frac{12}{1}$ and then multiplying $\frac{1}{3} \bullet \frac{12}{1}$ right straight across, top times the top and the bottom times the bottom, we get:

$$\frac{1}{3} \div \frac{1}{12} = \frac{1}{3} \bullet \frac{12}{1} = \frac{12}{3} = 4.$$

(b) $\frac{2}{3} \div \frac{1}{12}$ **is 8.** "Flipping over" $\frac{1}{12}$ (on the right) and then multiplying straight across gives the correct answer:

$$\frac{2}{3} \div \frac{1}{12} = \frac{2}{3} \cdot \frac{12}{1} = \frac{24}{3} = 8.$$

(c) $\frac{2}{3} \div \frac{1}{6}$ **is 4.** "Flipping over" $\frac{1}{6}$ (on the right) and then multiplying straight across gives the correct answer:

$$\frac{2}{3} \div \frac{1}{6} = \frac{2}{3} \cdot \frac{6}{1} = \frac{12}{3} = 4.$$

(d) $\frac{3}{4} \div \frac{3}{4}$ **is 1.** "Flipping over" $\frac{3}{4}$ (on the right) and then multiplying straight across gives the correct answer.

$$\frac{3}{4} \div \frac{3}{4} = \frac{3}{4} \cdot \frac{4}{3} = \frac{12}{12} = 1.$$

GO BACK TO CLASS. In math class, we use the words "take the reciprocal" instead of "flip over." So, in class, you will often hear, "To divide fractions, take the reciprocal of the fraction on the right and multiply." This is often abbreviated as "Multiply by the reciprocal." My only problem with this abbreviation is that students forget when to "multiply by the reciprocal" and they forget that they have to take the reciprocal of the right fraction, not the left fraction. Some students mistakenly use this procedure on multiplication problems, and some students "flip" the wrong fraction.

THE RIGHT HAND MOTION DIVIDES FRACTIONS. I believe (especially for you who are auditory learners or have verbal-linguistic intelligence) that if you are going to verbalize a math rule in your mind, it is best to learn it with all of the details. My preference is to learn this sentence: "To divide fractions, flip the guy on the right and multiply straight across." To cement this procedure in your mind, especially if you are a kinesthetic learner or have bodily-kinesthetic intelligence, repeat this sentence as you perform the following actions. Flip your *right* hand over, then hold two fingers straight out as you move your hand to the right. The two fingers remind you to multiply straight across—top times top and bottom times bottom. The actions accompanied by the detailed sentence spoken aloud will cement the procedure for dividing fractions into your mind. If you do the actions and say the sentence as you write out problems, the procedure becomes yours for life. You will eventually be able to do weird-looking algebra division problems without flinching.

TRY THESE EXAMPLES. Check with the solutions in the appendix.

1. $\frac{2}{3} \div \frac{2}{3}$ 2. $\frac{3}{4} \div \frac{1}{4}$ 3. $\frac{1}{4} \div \frac{1}{12}$

4. $\frac{5}{6} \div \frac{1}{12}$ 5. $\frac{1}{2} \div \frac{1}{6}$ 6. $\frac{4}{5} \div \frac{2}{10}$

12

Choosing Classrooms and Teachers

"Nothing is troublesome

that we do willingly."

THOMAS JEFFERSON

PARTICIPATE IN THE PROCESS

Choose to participate actively in the classroom environment. Getting to know your teacher and other students one-on-one increases your comfort level by giving you allies. By taking the lead in introductions, you could facilitate a positive, supportive classroom for others as well as for yourself.

Communicate with Your Instructor

When a student expresses good will, shows a desire to learn, puts energy into homework, studies, and comes to class, teachers are delighted. Teachers are a great resource and often go far beyond their contract requirements for their students.

Math teachers want to reach their students. They become teachers because they care about students, learning, and mathematics. They often blame themselves when students are not doing well. A few become defensive easily and feel that questions criticize them. A few have forgotten how difficult it is to learn math, and many of them have learning modes and combinations of intelligences that are different from their students'.

As department chair of my college math department, I received phone calls and visits from students complaining about or frustrated with their teachers. Usually I arranged for the student to talk to the teacher. Often this worked. It took good will and patience on everyone's part.

By communicating with your instructors, you give them necessary feedback and you help them understand your individual issues. After a one-on-one talk, both you and your instructor see each other's points of view a little better. It may be difficult for your instructor to listen to what you have to say, but it will be helpful, and almost all of them will be extremely grateful. If you need some support to speak with an instructor, math department chairs, deans, and counselors can give you ideas or might accompany you to assist the discussion.

If you have gone the extra mile to communicate with your instructor and it hasn't helped, attend class, tape the lectures, and get a tutor to help you. Do not let a poor teacher stand in your way of learning math. The math department chair, dean, or college counselors can help you find alternative resources.

"Fight to have a good teacher. Lobby for people who understand mathematical thinking. Talk to somebody (maybe a peer) who really is excited about or obsessed with math. They might be able to talk about it in a fun way even if they are at a higher math level."

—*Judy Schaftenaar, Ph.D., educator and administrator*

"Make the teacher teach you. Don't 'go at' the teacher and try to make an enemy or set yourself against them. But find out how you can get that teacher to teach you what you want to know.

"You have to take responsibility for your education. If you need to find out more about numbers or computers, you need to know that not every teacher is going to be able to teach you. You're going to have to find the right teacher or find the right teacher in the wrong teacher.

"I once had a teacher of art history. I loved his work. I talked to some of my fellow students and they found him dry and uninteresting. But I went to him after class. When I had a [subject] that interested me, I would bug him until he would sort of 'pour forth' the enormous knowledge that he had. I was taught because I demanded to be taught—not in a way to accuse him or to put him on the defensive but in a way that made him feel sympathetic and cheered by my interest. I learned a good deal from him.

"There are ways to treat teachers so that they will respond to you."

—*Elinor Peace Bailey, artist, doll creator, author, teacher*

Form a Study Group

Small-group study works well. Students become acquainted with classmates while they interact actively with the subject matter. As they talk, write, and do problems together in a group, they have the benefit of other minds as well as their own. They don't get stuck as often since they can frequently answer each other's questions. It is more fun struggling together.

Someone has to organize a study group or facilitate a support system within the classroom. Otherwise those things don't happen. Why not you? How would you do this? You could:

- Request that the teacher announce the forming of a study group and tell interested students to see you.
- Announce the group yourself giving times and place. The library, an empty classroom, the math study center, or the cafeteria are possibilities.
- Talk to a number of different students and suggest forming a group. Let those interested help you plan.

To make the group effective, develop an open policy that allows all questions. Encourage a supportive attitude and a positive tone. Invite others from your class to join you.

"Talk to people that really are excited about math. Part of getting stuck is not realizing the options you have. Instead of leaving, use that energy to resource something else. Change teachers if you have to."

—*Emily Meek, psychotherapist and artist*

"Talk to everyone you can about mathematics so that you can find somebody who can explain the concept to you in a way that you can understand it. Oftentimes your parents and teachers are not the best people to do that. Sadly, there are power struggles and emotional investment in both those relationships. Try other teachers who don't have an emotional investment in your grade, brothers, sisters, adult friends of the family . . . By mentioning the problem over and over and over again, eventually you will find somebody who can understand."

—*Sarah Kershaw, student and math tutor*

To ensure her success as a student, Sarah says, "If the teacher is somebody I can learn from, I attend the class religiously. If the teacher is somebody who is confusing me or is condescending, I find a way out of that class, whether it's taking it and not attending or it's transferring classes. This is my education. This is my time and I have a right to use it effectively."

RECOGNIZE SAFE AND THREATENING LEARNING ENVIRONMENTS

Unfortunately, some math classrooms and tutoring sessions have an almost paralyzing environment of fear. Brain scientists know that threat causes "emotional flooding" in the brain and temporarily renders the brain biologically and chemically unable to learn.

You can recognize the difference between a safe learning situation and a threatening one by how questions are handled.

• In a safe learning environment, all questions are acceptable but time constraints are a reality. In classrooms, students often have to get their answers outside lecture because of limited class time. As a student, ask all the questions you have and let the teacher or tutor judge the time constraints. The worst thing that can happen is that your question won't get answered the first time you ask it. Eventually you will get an answer.

• In a threatening environment, questions may be judged unworthy. This is *not* acceptable. There are no unworthy questions—no bad questions—no stupid questions. Questions are requests for information that students do not have and do not know. If you do not know something, is that not the reason that you are asking? When you don't know something, you don't know it.

Students have a right to ask all their questions. At the same time, the teacher or tutor has a right *not* to answer all of them because of time limitations. Judging students' questions is no one's right.

Even as you concentrate, you still won't catch everything. When the material is mostly new, much of what is said goes by you. That is the human condition. That is normal. Your brain needs processing time to integrate information with previous ideas. Asking questions helps you to sort, process, clarify, integrate, and understand.

When questions are not asked in a classroom, that doesn't mean everyone understands. It can mean that people are so confused they don't know where to start. If you ask your question, you might just clarify the session for others as well as for yourself.

There are many places to get help besides from your teacher or tutor. Your resources include fellow students, your book, other math teachers, other tutors, math students standing in the hallways, math tutoring centers, and other textbooks from the library. Getting acquainted with resources when you are calm and relaxed helps you more easily find them later on when you're stressed and urgently need help.

If you find a tutor's actions or unwillingness to answer your questions not helpful, choose another tutor. Some tutors are trained and some are not. Some tutors are good and some are not. It is O.K. to fire your math tutor and find another.

"I wish I had known that there are just plain unskillful teachers that may be different from you, the student, trying to understand the math. Unskillful teachers have kind of an army approach of 'I got my knuckles rapped with a ruler and I had to memorize everything. I did it and that was good enough for me. That's how you're going to do it.' [Students need to know] that there are other teachers or programs."

—*Emily Meek, psychotherapist and artist*

"I think I could have been much more successful in algebra if it had been approached differently and I had not been intimidated by 'math' and the grading system. I know now that failures in school, grade-wise, are not what we take them to be when we're in school."

—*Bob White, musician, artist and retired police officer*

"I wish I had known someone who was willing to think—to take me seriously as a mind and not trivialize me because I don't think in the way that they think. I wish I'd known that there are such people."

—*Elinor Peace Bailey, artist, doll creator, author, teacher*

The Ideal Math Classroom

"A good teacher is the holiest of God's creatures. I don't think that there's another role in our culture that deserves the rank of holy except a great teacher who excites you. And the bad teachers should be made to march in chains."

—*Barbara Sher* (McMeekin, 2000)

I love what Barbara Sher says above about teachers. However, I know that whether I would be ranked holy or marched in chains would depend on which of my students you talked to. Learners and teachers are all individuals and different from each other. As a math student, remember that there are no perfect teachers and no perfect classrooms. There are, however, teachers and classrooms that can work for you.

Evaluate your math class and math teacher. If you can recognize a positive classroom environment, you may be able to help create one. It is too easy to blame the instructor for frustration with math. Total honesty and introspection pinpoint where the problem lies. The bottom line is that since no one can change others, the real power lies in changing ourselves.

It is *you* who must make your math experience worth your time and effort. If the problem is not you, you might be able to facilitate your learning in a negative environment by observing and not absorbing or personalizing the negativity. You cannot make your teacher change, but you can evaluate the class atmosphere.

Below I quote the math department philosophy from my college. As I reread this teaching philosophy, I recognize how "ideal" it is. Being human, I know that, as a math teacher, I can only aspire to these goals. I know that there are days when I am distracted or under the weather or just plain insensitive. As hard as I try to keep my classrooms always safe, encouraging, interesting, and accessible, I know this will not always happen.

As you take charge of your learning experience, you will be able to forgive your math teachers for not being perfect, because you will know how to get what you need.

Santa Ana College Math Department Philosophy

When the term "math anxiety" was coined by Sheila Tobias in the 1970s, a name was given to a phenomenon that was rampant in this country. Having a name for this phenomenon or set of symptoms gave math students a way to separate themselves from their experiences and to take control.

In this math department, we believe in empowering students. Negative past experiences, learning disabilities, and current life stressors all affect a student's ability to gain access to the linear, analytic functions of the brain required to do math.

It is now widely known that Albert Einstein and Thomas Edison appeared dull and slow as students. Winston Churchill flunked English. Leonardo da Vinci, Ludwig von Beethoven, Louis Pasteur, and Hans Christian Andersen had learning disabilities. As we look out over our math students or grade their exams, we cannot know the depths of their abilities. All we know is what they can currently access.

Therefore, it is in their best interests that we provide an atmosphere that is safe and positive so that they can begin to open their minds to math. This is not to say that we "lower our standards" or that we become floor mats and "water down our courses."

It is to say that we mirror positiveness and possibilities to them. We provide them with support. We give them consistent feedback on the bits of progress that they make so that they continue to put one foot ahead of the other working their way up the math mountain.

(continued)

Santa Ana College Continued

We may be the first math teacher they ever had who believed that they could do math or the first to present it in enough different learning modes so that they could finally grasp it. We may be the first math teacher who ever gave them permission to make mistakes and to take the risks that allow them to learn.

- When we as math teachers are willing to examine the shadowy parts of our academic past and think about the courses we enjoyed the least,
- When we are willing to recognize that our math abilities gave us a certain intellectual status so that we had permission to not do so well in perhaps P.E. or English comp,
- When we are willing to admit our discipline is no better and no worse than any other academic discipline but that it currently enjoys a reputation as being the best indicator of intelligence,

then we can truly realize the incredible courage it takes for students whose skills lie elsewhere to enter our math classrooms.

Therefore it is our belief in this math department that to be truly effective with our students, we need to recognize the possibilities that are keeping our students from learning. We need to encourage, encourage, and encourage. We need to facilitate our students' use of the extra supports that we have on our campuses for tutoring, coping with math anxiety, personal counseling, and diagnosing and coping with learning disabilities. It is also helpful if we have read materials on math anxiety so that we do not perpetuate some of the negative ideas that fill students' heads and cause static preventing clear thinking.

Being a math teacher

- who clearly verbalizes expectations and ground rules—writing them out in our course outlines,
- who gives class presentations that are well thought out and organized,
- who calls students by name and actively engages them positively in the learning experience,
- who varies classroom activities to accommodate diverse learning modes and attention spans, and
- who is knowledgeable about support services and encourages students to use them,

can go a long way toward reducing math anxiety and releasing student energy to be used on math.

We have a challenging and rewarding job to do. Isn't it wonderful?

Pushing Your Limits

CHAPTER

1. Write in your journal about how you can or do actively participate in your math classroom. What new possibilities have occurred to you while reading this chapter? Make a plan to get what you need in a math classroom.

2. Consider forming a study group. Make a plan or ask your instructor to help you. Write down your plan and the advantages of forming a group.

3. Which of the people quoted in this chapter spoke for you? How are you alike? How are you different? Record these thoughts.

4. What is your reaction to the Math Department Philosophy? You may wish to copy this philosophy and share it with your teacher or fellow students.

5. Mastering Math's Mysteries, Chapter 12, will give you practice using four common math words with your fraction skills from Chapters 7–11. You will also get to see some creative memory techniques.

Sum, Difference, Product, and Quotient

LEARN THE VOCABULARY: SUM, DIFFERENCE, PRODUCT, AND QUOTIENT.
These four words are commonly used for operations in mathematics.

Sum means to add.

Difference means to subtract.

Product means to multiply.

Quotient means to divide.

The four operations (add, subtract, multiply, and divide) are usually learned in that order. If you remember the words *sum, difference, product,* and *quotient* in the same order, you are likely to recall their meaning. Below are some weird sentences, some visual cues, and a song to help you remember the meanings of these four words that show up everywhere in math.

S D P Q

Some Do Problems Quickly.

Speed Doesn't Promise Quality.

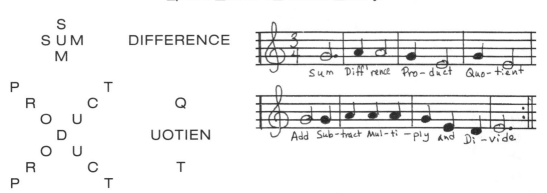

Complete these sentences:

1. The sum of 8 and 3 is _____.
2. The sum of 10 and 12 is _____.
3. The difference of 9 and 5 is _____.
4. The difference of 18 and 7 is _____.
5. The product of 4 and 9 is _____.
6. The product of 8 and 3 is _____.
7. The quotient of 36 and 6 is _____.
8. The quotient of 27 and 9 is _____.

Complete these sentences:

9. The product of 3 and 7 is _____.
10. The sum of 8 and 9 is _____.
11. The quotient of 3 and 7 is _____.
12. The difference of 7 and 3 is _____.
13. The quotient of 24 and 6 is _____.
14. The product of 8 and 9 is _____.
15. The difference of 11 and 4 is _____.
16. The sum of 7 and 14 is _____.

PUT IT TOGETHER. Practice vocabulary and fractions.

17. Find the sum of $\frac{4}{15}$ and $\frac{9}{15}$.
18. Find the quotient of $\frac{2}{3}$ and $\frac{1}{3}$.
19. Find the product of $\frac{3}{8}$ and $\frac{1}{3}$.
20. Find the difference of $\frac{7}{12}$ and $\frac{2}{12}$.

13

Making Math Memories

**KINSEY MILHONE
IN SUE GRAFTON'S
"O" IS FOR OUTLAW**

"I used to imagine I could hold it all in my head, but memory has a way of pruning and deleting, eliminating anything that doesn't seem relevant at the moment."

Remembering is not about memorizing. It is about making strong, long-lasting, and accessible brain connections—laying down enduring pathways.

Healthy humans possess good memories. They remember events, facts, and details. They just don't necessarily remember all that they wish to remember. This chapter explains how you can consciously improve your recall of what you need for math.

Math students often believe they must memorize everything. This is not true. Learning math is about making brain connections. This involves understanding and noticing patterns and then practicing in as many different ways as possible.

You want those connections to last long term. They store your math vocabulary, concepts, and procedures. It is those long-term connections that assist you in making future mathematical insights and solving problems.

Your capacity for long-term brain connections is practically unlimited. This means that you have plenty of space to store mathematical ideas. However, those pathways rearrange themselves, update with new information, and deteriorate constantly. This means that your math memories get changed and confused unless they are *made carefully* and *reinforced by constant practice*.

This chapter will follow the three-step process for forming long-term connections and give you activities for each step of the way so that your long-term math memories will be lasting and accessible.

STEP 1. HOW TO TAKE MATH INTO YOUR MIND TO REMEMBER IT WELL

Make strong connections to start! Here are seven techniques to do just that. These activities will help you form solid long-term math pathways in your mind. Read each suggestion and check the ones you plan to use soon.

Decide What You Need to Remember and CHOOSE to Remember It

Class notes and chapter review problems can tell you what is important. Highlight important facts. Make a record of what you have assessed as important to remember. This record could be in your notes, in your journal, or in the form of a web. (See the next chapter for an explanation of webbing.) To review what is important, summarize your choices on index cards. Write a problem or vocabulary word on one side of a card and the solution or definition on the other side.

Have FUN

Seriousness does not help brain function. Intention and good humor do. When you play with your work, you will do better work.

Make jokes and laugh a lot. Sometimes just enjoy the absurdity of it all. Work with positive people in a study group. Don't sit by negative complainers. Math jokes are usually corny. Even "corn" can give you a good time.

Math Corn

Q: What did the little acorn say when he grew up?

A: Gee-Ah'm-A-Tree

Get FEEDBACK Quickly

Use what you learn immediately to clarify that it is correct. You may have misunderstood and formed incorrect connections in your mind. The more you travel the wrong brain pathway by practicing incorrectly, the more difficult it is to correct. Here are two examples of how to get feedback. An entire list of feedback activities can be found in Chapter 10. For example:

- Work three or four homework problems, then compare your answers with those in the back of the book.
- Teach what you learned to another math student in your class.

OBSERVE Math Closely for Details

Here are some ways to do that.

NOTICE SIGHTS, SOUNDS, OR SMELLS. Even math classes change from day to day.

- The equations you solved the day the teacher wore a Hawaiian shirt could be recalled as the "Surfer Problems."
- The Quadratic Formula could be remembered as what you learned on the day your neighbor wore that really smelly perfume.

PICK OUT KEY WORDS. If you notice key words and learn their meaning, you will organize your overall knowledge. For example, nearly all beginning algebra problems can be categorized as *simplifying, solving, graphing, factoring,* or *evaluating* problems. When you see those words in directions, you will already have a category of problem types in your mind with which to associate.

LOOK FOR PATTERNS. For example:

- Any number that divides evenly (with no remainder) by the number 5 ends with a five or a zero. The numbers 510, 275, and 1,365 can be divided evenly by 5. The numbers 712, 1,478, and 653 cannot.
- All equations have two sides—a left side and a right side—separated by an equal sign. $3 + 2 = 5$ is an equation. $3 + 2$ is the left side and 5 is the right.

WRITE OR DRAW WHAT YOU SEE AND THINK.
Visualize it and say it in your own words. Find
unusual, vivid relationships between nonmathe-
matical ideas and mathematical ideas. For
example, the number "pi" or π is used with circles to find the circumference (a measure of the
distance around the outside of the circle) and
the area (a measure of the surface inside the circle). To remember "pi" think of the fact that
circles are round like pies.

COMPARE FOR SIMILARITIES AND DIFFERENCES.
To learn something new, relate it to things you
already know, and look for ways it is the same or
different.

- -8^2 and $(-8)^2$ look very similar, but -8^2 is -64 and the other, $(-8)^2$, **is** 64.
  That is the difference between "owing $64" and "having $64."
- The formula for area of the square, rectangle, and parallelogram (which
  are all four-sided) is base times height. The formula for the area of
  another four-sided figure, the trapezoid, is also base times height *except*
  that you have to find the average of the trapezoid's two bases (because
  they are different), like this:

area of trapezoid = average length of bases times height = $\left(\dfrac{b_1 + b_2}{2}\right) \bullet h$

CONSIDER THE CONTEXT. Say to yourself. Where does this idea fit into a larger picture? What are the circumstances? What does this idea mean here? For
example, .02 + .8 + 2.35 looks like money, even if it's inches. You know how to
add money!

Be ACTIVE with What You Want to Learn

Manipulate and experiment. Find real-life examples of what you are learning.
Make analogies that you understand. Talk about it, read more about it, sketch
it, question it, dance with it, rhyme with it, web it . . . For example:

- Think about distances, rates, and times as you drive, or pretend your
  eraser and a paper clip are the two cars racing in the problem. Replicate
  the distances, speeds, and times with actions.
- Notice geometric shapes and patterns in nature around you.
- Use paper or cookie dough to make models of problems.
- Work problems on a chalkboard.
- Use the Finger Technique to learn the nine multiplication tables.

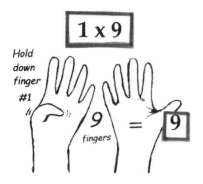

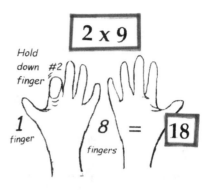

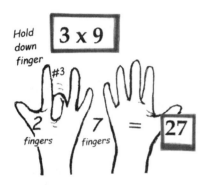

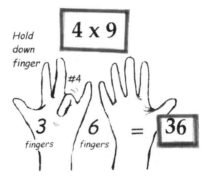

CHUNK the Information

Break information down into smaller, easier-to-remember pieces of three to four bits each. This makes it more manageable. For example, to learn the multiplication tables, you may already know the twos, threes, fours, fives, and nines. (Use the Finger Technique for nines.) All that is left to know is 6 • 6, 6 • 7, 6 • 8, 7 • 7, 7 • 8, and 8 • 8. Just work on these six facts, three or four of them at a time, and you've got it. Notice that three of these facts involve square numbers.

6 • 6 = 36 (a square number!)	7 • 7 = 49 (a square number)
6 • 7 = 42	7 • 8 = 56
6 • 8 = 48	8 • 8 = 64 (a square number)

Tune Everything Else OUT

Avoid interference. Study where your brain gets the privacy it needs. You will discover and know what works best for you if you experiment.

- *For auditory learners:* Use ear plugs or a walkman. Turn down the TV. Stake out a quiet section of the library or the math-tutoring center.
- *For visual learners:* Clear the visual field in front of you. In class, sit in front so you don't see other students and non-math action. Use a study carrel in the library. Clear your worktable or desk before you begin. Choose a desk or work area facing a neat and pleasing background.
- *For kinesthetic learners:* Choose workplaces where you have space to move as you need. Give thought to your physical comfort.

STEP 2. HOW TO STORE MATH WELL IN YOUR MIND

Keep those connections! Once you take math processes and ideas into your mind, these six techniques will help your brain to store them. As you read, check those actions you are ready to take.

Allow Yourself SETTLING Time

Pace yourself. Your brain needs to process math information with no new challenging input coming in. Spend from 20 to 50 minutes learning and then spend about 10 to 15 minutes resting or doing something else unrelated to what you just learned. Settling time does not have to last long. It can be as short as walking to the next class, taking a restroom break, chatting briefly with your neighbor, or stepping outside to get a breath of fresh air. During a math class that lasts over an hour and a half, those breaks are extremely important. A knowledgeable teacher will build settling time into the class structure. Taking a break to joke around or move the class into groups or to the board allows students the breather they need. I use math songs and relaxation exercises to provide settling time in my classes.

USE It Not to Lose It

Repetition and practice reinforce the wiring in your brain.

- Revisit the concepts and processes within hours, then again days, weeks, and months later.
- Talk to yourself about math and discuss math ideas with other students. Tutor others. Work problems over again without the aid of your book or notes.
- Work the chapter review problems from the book. Make a list of examples that your teacher presented in class. Work those problems once a week throughout the semester.
- Besides doing the current homework, do three different review problems each day.

SLEEP

You require sufficient rest, especially early morning sleep. The hours between 3 A.M. and 6 A.M. are the hours you process and understand your long-term memories during REM (rapid eye movement) sleep. Studies show that a lack of sleep interferes with brain function.

Say to Yourself: "This information is IMPORTANT"

Nonessential information disappears quickly.

Use Hard Storage for SUPPORT

Write things down. Use a personal calendar, a notepad, note cards, a journal, sticky notes, or computer files to record and recall essential information. The next chapter will show you how to use webs to form a visual summary of ideas about a concept.

Take careful notes in your math class. Record all assignments. Copy all of what the instructor writes on the board as well as verbal explanations. Develop a note-taking system so you can find assignments, important points, or examples at a glance.

As you study, write down key concepts with problems that exemplify those ideas.

Develop HABITS

Your habits will form well-established pathways in your brain to be activated by exam problems. For example, form the habit of taking a deep breath every time you begin to work on a word problem. Read the word problem through once just to figure out what you are requested to find. Take a second deep breath, expelling all the air. Then read the problem through two more times before you ever begin working. These habitual activities will relax you and remind you that you are not expected to understand word problems immediately.

As another example, I teach my students a four-step strategy for solving algebra word problems. Making this four-step strategy a habit gives them a plan of attack for any algebra word problems they encounter. The four steps are:

- Write specifically what you are asked to *find* in the problem.
- Explain what every *letter represents* that you substitute for the number(s).
- Show the algebra *work* that solves the problem.
- Write the *answer in a sentence* and *check* this solution in the original problem.

STEP 3. HOW TO RETRIEVE THE MATH YOU HAVE STORED IN YOUR MIND

Find those connections! Each step of making long-term memories is crucial. If the pathways have not been formed and stored well, they cannot be retrieved or remembered.

The key to retrieving or accessing what you have stored in your brain is *practice*. The more you practice the math concepts and techniques, the more easily you can remember them as needed. Here are three techniques that you can consciously use to recall facts, formulas, procedures, or definitions. These techniques don't explain why a procedure works, but they do help you recall how to do the procedure once you understand it. Understanding also helps your recall.

LINK Words or Ideas with Actions

Actions create motor memory. Move around as you say the ideas aloud. Create a little dance with them. For example:

- Flip your right hand over and say the word "Reciprocal." The reciprocal of $\frac{2}{3}$ is $\frac{3}{2}$. The reciprocal of 5 is $\frac{1}{5}$ because 5 is $\frac{1}{5}$.
- To remember how to multiply and divide fractions using actions with your right hand, see Mastering Math's Mysteries, Chapters 10 and 11.

Linking words or ideas with actions can be used at any math level from basic math into more advanced math. One of my colleagues had her trigonometry students line dance the shapes of the six basic trigonometric graphs. A lot of jumping and slinking up and down can make those graphs easily memorable.

I memorized a French poem while walking back and forth to class in 1963. Unfortunately, almost 40 years later, this poem is the only French I remember from three years of studying the language in college. I do not even know what the words mean, but they are inscribed into my brain. The only explanation I have is that the motion formed strong connections in my mind. Hindsight is great. Now I know I could have remembered much more by reciting my French lessons aloud as I walked from my dormitory to class! Although I am a visual learner and spell fairly well in English, my recollection of this poem is an *auditory memory*. If I were to record my version of the poem it would be filled with misspellings galore, but the "sounds" would match!

LINK Words or Ideas with Images

Vivid pictures involve sensory memory.

- Disneyland knows this technique. They place images of Disney characters on signs throughout their parking lots to catch the eye of park goers and remind them of the locations of their cars. (Can you imagine searching the Disneyland parking lot for your lost vehicle?)
- Earlier, I showed you a four-step process I attempt to get my algebra students to use habitually. On exams, I remind them of this technique by putting the four dots below where they will work their word problems on exams. Each dot serves as a visual cue for a step in solving their problem:
  - •
  - •
  - •
  - •
- Writing the important algebra words *sum, difference, product,* and *quotient* in the form shown in Mastering Math's Mysteries, Chapter 12, can help you remember what to do when you see those words.

MAKE UP Sentences, Goofy Words, Songs, and Rhymes

The more meaning they have for you, the better you will remember. Making them funny helps too.

- The sentence "**P**lease **E**xcuse **M**y **D**ear **A**unt **S**ally" recalls the order of operations for basic math students: **P**arentheses, **E**xponents, **M**ultiplication and **D**ivision (from the left to the right), and **A**ddition and **S**ubtraction (from the left to the right). Sometimes I say, "**P**lease **E**xcuse **D**ear **M**y **S**ally **A**unt" to remind students that the operations multiplication and division are done from the left to the right and sometimes multiplication is first and sometimes division is first. The same with addition and subtraction.

$25 \div 5 \bullet 4 - 3(4 - 2) + 5^2$	Work inside <u>P</u>arenthesis. $4 - 2$ is 2.
$= 25 \div 5 \bullet 4 - 3(2) + 5^2$	Do <u>E</u>xponents. 5^2 is 25.
$25 \div 5 \bullet 4 - 3 \bullet 2 + 25$	Do <u>D</u>ivision. It is *left* of the Multiplications.
$= 5 \bullet 4 - 3 \bullet 2 + 25$	Do <u>M</u>ultiplications *left to right.*
$= 20 - 6 + 25$	Do <u>S</u>ubtraction. It is *left* of Addition.
$= 14 + 25$	Finally, <u>A</u>dd.
$= 39$	

- Earlier, you read about recognizing key words and learned that the five most important procedures taught in beginning algebra are: simplify, solve, graph, factor, and evaluate. The following words (by Mike Petyo and Cheryl Ooten) sung to the Christmas carol "O Come All Ye Faithful," help my students remember those five procedures, which show up in most of their algebra problems. Knowing those five words, what they mean, and some examples of how to do each procedure can give my students a nice summary of what they have done in one semester's work.

  ### I Love My Algebra

  I love my algebra.

  Morning, noon, and night,

  I do algebra whenever there is light.

  Simplify and solve all that I can take.

  Graph, factor, an' evaluate.

  Graph, factor, an' evaluate.

  Graph, factor, an' evaluate,

  They're a piece of cake.

- Here is an example of this technique from a more advanced math course, trigonometry. *Even if the math makes no sense to you, you can appreciate the concept of using a goofy word to memorize some math ideas.* In "Chief **SohCahToa**," each letter of the chief's name recalls how to use the lengths of sides of right triangles to find basic trigonometric functions or relationships with angles:

  Sine of an angle equals the Opposite side divided by the Hypotenuse.

  Cosine of an angle equals the Adjacent side divided by the Hypotenuse.

  Tangent of an angle equals the Opposite side divided by the Adjacent side.

- Another trigonometry example: Jaime Escalante, East Los Angeles math teacher featured in the movie *Stand and Deliver,* taught his high school students "All Seniors Turn Crazy" to help them remember which trigonometric functions are positive in the four quadrants of the xy-axis. *If you don't understand this math, just chuckle at the sentence that he used to help them remember the idea:* In Quadrant I, All functions are positive. In Quadrant II, the Sine function and its reciprocal, the cosecant, are positive. In Quadrant III, the Tangent function and its reciprocal, the cotangent, are positive. In Quadrant IV, the Cosine and its reciprocal, the secant, are positive.

In summary . . .

To influence what you remember and make strong, long-lasting, and accessible brain connections:

- Be active and have fun when you learn.
- Choose what is important and consciously convert it to memory.
- Use it or lose it.
- Employ actions, images, songs, rhymes, or sentences for recall.

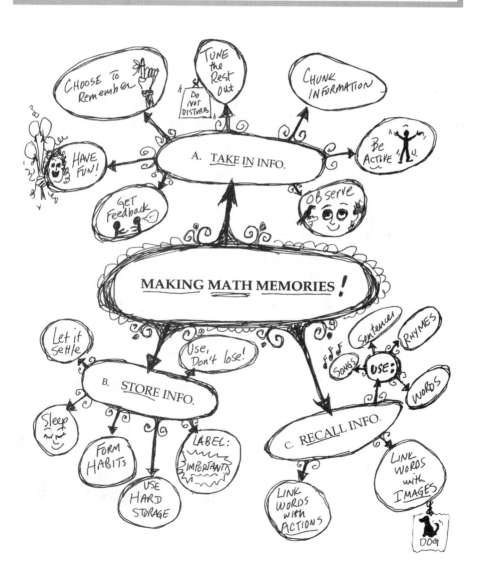

JUST DO IT

The title and inspiration for this book came from a song I made up after listening to a song about the "mean [something I don't remember] blues" on the radio. I substituted the word "math" for the word in the middle and I was off with a new song and new idea. Try it. It doesn't have to be great music to be fun or helpful.

Doll creator Elinor Peace Bailey told me that she tells her students who want to invent a new doll that they have to be willing to first just make a doll—any doll—a bad doll or whatever, any way that they can. In other words, she says, "Just do it."

So I say, just make up some little tunes or rhymes with words you want to remember. Just do whatever it takes.

Pushing Your Limits

1. Talk to people about memory skills. Ask them if they have any techniques that have worked for them. Talk about what you remember well and hypothesize about why that is. Record your insights.

2. Write down three ways that you can improve your recall. Make up personal examples for those techniques.

3. Make up a silly word or sentence to recall an important fact in the math that you are studying right now. Be outrageous. Make up a song!

4. Practice matching images with words by drawing an appropriate image by some of the points in the web on the previous page. Then draw those same images by the points within the chapter. Actually copying the web into your journal with your images will help you remember the techniques in this chapter for making math pathways!

5. Practice fraction operations again in Mastering Math's Mysteries, Chapter 13. The song/poem is a memory device. Review Mastering Math's Mysteries, Chapter 12, to see image, song, and nonsense used for making solid math memories.

Memory Devices for Fractions

SING IT OUT. The next page shows a song/poem about adding, subtracting, multiplying, and dividing fractions. It is a memory device, but it also can instruct you. Do the examples at the bottom of this page by reading or singing the song and looking at the examples that are worked out. The only part of the song that hasn't been explained in Chapters 7 through 11 is how to reduce a fraction at the end. Perhaps you can figure that out by looking at the song and the example. Check your answers with those in the Appendix.

WHO IS X? We also haven't discussed the letter "x," used frequently in math. I call it "the mathematician's favorite letter." Just remember that it is a dummy stand-in for a number. "x + 3" means "a number plus 3." "3x" means "3 times a number."

Try these:

1. $\dfrac{2}{7} + \dfrac{4}{7}$

2. $\dfrac{8}{x} - \dfrac{3}{x}$

3. $\dfrac{x}{9} + \dfrac{3}{9}$

4. $\dfrac{16}{y} - \dfrac{4}{y}$

5. $\dfrac{3}{7} - \dfrac{x}{7}$

6. $\dfrac{5}{14z} + \dfrac{x}{14z}$

7. $\dfrac{2}{7} \cdot \dfrac{4}{7}$

8. $\dfrac{3}{4} \cdot \dfrac{x}{7}$

9. $\dfrac{4}{5} \cdot \dfrac{2}{3}$

10. $\dfrac{1}{2} \cdot \dfrac{7}{y}$

11. $\dfrac{3}{5} \cdot \dfrac{6}{5}$

12. $\dfrac{3}{7} \cdot \dfrac{3}{5}$

13. $\dfrac{1}{5} \div \dfrac{4}{7}$

14. $\dfrac{2}{5} \div \dfrac{3}{x}$

15. $\dfrac{4}{3} \div \dfrac{3}{1}$

16. $\dfrac{5}{7} \div \dfrac{1}{7}$

17. $\dfrac{5}{6} \div \dfrac{2}{3}$

18. $\dfrac{1}{3} \div \dfrac{1}{3}$

Fractured by Fractions—NOT

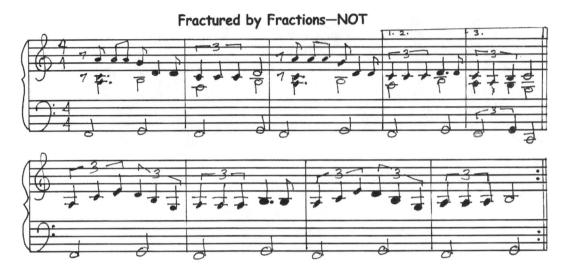

Fractured by Fractions—
I thought I was.
Now I can do them,
It seems, just because—

For adding, subtracting,
The bottoms must match.
So keep them and change
   the tops.
That is the catch.

$$\frac{4}{10} + \frac{3}{10} = \frac{7}{10}$$

$$\frac{x}{6} - \frac{7}{6} = \frac{x-7}{6}$$

Mul-ti-ply fractions
Right straight across.
It's cool that fi-nal-ly
I'm "Fraction Boss."

$$\frac{2}{5} \cdot \frac{3}{7} = \frac{6}{35}$$

Di-vi-ding—you flip the guy
On the right side.
Straight across—you're done
When you've multiplied.

$$\frac{5}{7} \div \frac{3}{4} = \frac{5}{7} \cdot \frac{4}{3} = \frac{20}{21}$$

I'll fracture fractions—
Accept no excuse.
Just factor, cancel
The end to reduce.

$$\frac{24}{36} = \frac{2 \cdot 2 \cdot 2 \cdot 3}{2 \cdot 2 \cdot 3 \cdot 3} = \frac{2}{3}$$

14 Webbing

"Every individual is a

marvel of unknown and

unrealized possibilities."

GOETHE

Nottawa Woods
11/99
C. Dexter

WEBBING

Webbing is a powerful tool for taking notes, reviewing math, preparing for exams, and planning your studies. Also called graphic organizers, clusters, concept maps, Mindmaps®, and sequential organizers, webs increase math understanding by connecting and organizing concepts.

You have seen simple webs summarizing ideas in Chapters 5, 6, 7, 8, and 13. Webs take you beyond linear thinking. They clarify ideas associated with a basic focus. They illuminate connections and order, and they make ideas accessible.

The skills involved appear in the journals of some of the world's most inventive minds, such as Leonardo da Vinci, Thomas Edison, Albert Einstein, and Charles Darwin. Author and educator Tony Buzan (1994) linked these

How to Web

1. In the center of a blank sheet of paper, write the main idea or focus you wish to explore. Preferably use one word or a symbol. Imagine that this is the hub of a bicycle wheel.

2. Surround the main idea with all of the related ideas by drawing lines out in all directions like the spokes of a wheel. Print one word or symbol representing each related idea on the line or at the end of it.

3. Elaborate on related ideas. If you wish, add spokes to the related idea making a mini-wheel.

4. Add color. Show any other connections. Use pictures, words, and symbols.

skills to brain function and popularized the webbing technique through training and writing. He says that the results of this process mirror the associative thought processes of the brain. His book *The Mind Map Book* contains detailed examples and many exciting applications.

Buzan says, "The only barrier to the expression and application of all our mental skills is our knowledge of how to access them" (p. 33). Webbing can help you access more of your mental skills.

Be nonjudgmental as you make your webs. Your lines do not have to be perfectly straight and your pictures do not have to be masterpieces. You can use any color and any color code that appeals to you and suits your thinking.

This process is a tool for you to use. As you experiment with it, you will find what works best for you. Be willing to explore. Remember that everyone thinks differently, so your webs will not be the same as those of others. You can, however, get ideas from other people's webs to improve your own. If you are uncomfortable using this technique, you could potentially break through to a *whole new way of thinking* by persevering and using it.

Your webs are for you alone. There is no right or wrong way to make them. You can edit them as you notice new connections. You can even use the same main focus and start over making a whole new map.

Redrawing your web several times, adding details each time, helps you learn the ideas that you want to fix in your mind. Redrawing it is especially helpful when you prepare for an exam, review your course, set the ideas of the course in your mind, or look for new connections in the material. If you need to make a presentation in your math course (or any of your other courses), webbing the material and then redrawing your map will cement the presentation in your mind. You can even sketch your map on your exam or for your audience to recall details.

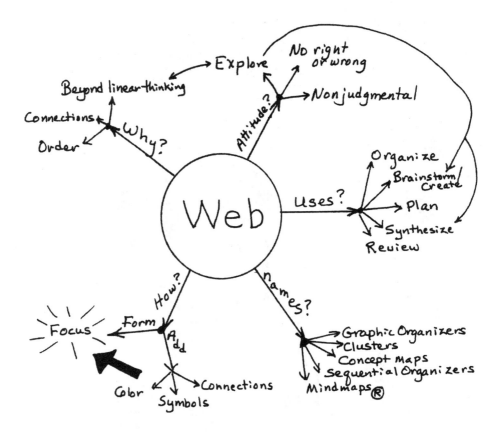

SUGGESTIONS FOR WEBBING WITH MATH

1. TAKE NOTES. Listening carefully to your math instructor for the organization of the lecture, put the topic of the class day in the center of your paper and fill in different procedures and examples as related ideas. You can circle related ideas and draw in arrows to show connections. Here is a web of notes on adding fractions. Remember that your thought process is unique, so this web may make no sense to you. Your web would look very different.

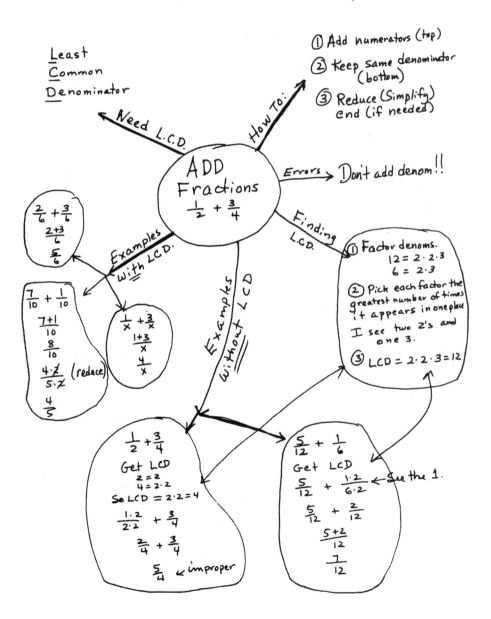

2. REVIEW YOUR MATH COURSE. Put the name of your course or concept you want to review in the center as the hub. Add the related ideas surrounding it like the spokes of a bicycle wheel. Here is an example of using this technique to review basic algebra. You do not have to understand all of this web to notice that I consider five major ideas in algebra to be: simplify, evaluate, factor, graph, and solve.

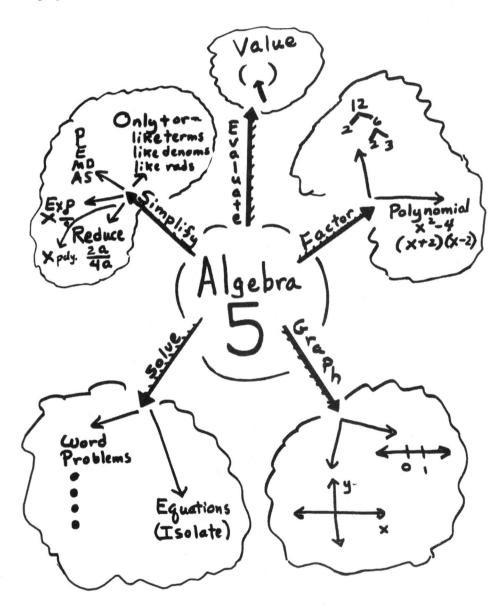

3. PREPARE FOR EXAMS. Several days before a test, write the words "Next Exam" or "Exam #3" or "Chapter Four Exam" in the center of a blank paper. Using your notes and your book, find the main topics that will be covered. Print these topics on the spokes and fill in words and examples appropriately. Here is an example of a web for preparing for an exam that could be given early in a beginning algebra course. Doing the sample problems in the web taken from class notes would be excellent test preparation.

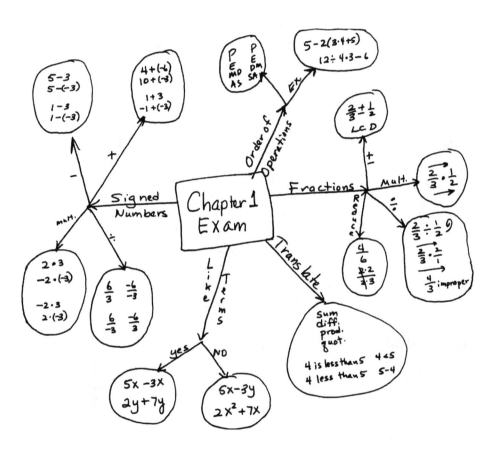

4. PLAN YOUR STUDIES. As the main focus for planning, you could write "To Do," "Study Activities," "Week's Work," "Final Exam Prep," or whatever you need to plan. Brainstorm activities that need to be done as the spokes and then break each activity down further. Draw in connections. Use highlighters to mark your priorities. Here is an example of a map for "Study Activities."

MY WEBBING EXPERIENCE

I had always considered myself a noncreative, linear thinker until I learned to web, observing Santa Ana College counselor Dennis Gilmour as he taught my math students in his counseling class. Dennis assigned our students to make webs about themselves as a preliminary step in developing autobiographies for scholarship applications. As I experimented with my own webs, I was able to get my random thoughts and ideas down more quickly. Dennis required our students to use colored pencils, so I bought some too.

After completing a web in color one weekend, I looked out my window to rest my eyes and saw an awesome sunset over the mountain crest. I remembered my childhood box of four hard paints that limited and frustrated me so in trying to get a sunset down on paper when I was eight years old. It occurred to me that my box of 32 colored pencils had many shades of yellow, pink, purple, and green. Excited and feeling creative and adventuresome because of my webs, I began to make colored marks on the page, making my first serious drawing, which led to the drawings you see at the beginning of each of these chapters.

To create this book, I made new webs constantly as I worked on the manuscript. When I felt stuck or inspired, I would take five or ten minutes to brainstorm all of the ideas I thought should be in the book. Often I started my daily writing by webbing my ideas for the day's topic. I put the main idea in the center of my journal page and, with stream of consciousness, surrounded it with all of my thoughts about the topic. Then I put lines and arrows between related ideas. Often I would look back at previous webs on the same topic to see if I had new ideas or connections. Writing came more easily after this activity.

Pushing Your Limits

14

CHAPTER

1. Web by writing the words "Math Success" in the middle of a big sheet of blank paper. In five minutes, surround these words with the actions you now associate with Math Success.

2. Web for test prep. Write down the words "Next Test" in the middle of a big sheet of paper. In five minutes surround these words with the concepts and problems that you believe will be on your next test. Take 10 more minutes to go through your book or class notes, adding and elaborating. Put page numbers or problems by topics. To continue your test preparation, set aside 30 minutes to write down specific problems for each topic. Copy the answers on a separate paper. Set aside another 30 minutes to work the problems and correct them. If you are not currently in a math class, do this exercise for any of your classes.

3. Copy the webs from this book that are relevant to your math study. Reorganize them to fit your own way of thinking.

4. Do Mastering Math's Mysteries, Chapter 14, to continue webbing practice with math.

Webbing

Make a web focusing on fractions. Put the word "Fractions" in the middle of a blank piece of paper. From that focus, branch out. You can use the four operations and examples. You can use symbols and pictures. You can show fractions and use color. Show your web to your instructor, a tutor, or a fellow student so that he or she can suggest any examples or additional subtopics that you may wish to add. Remember that it is your web, but that someone else's suggestions might make it more useful for your understanding of this topic.

If you are in algebra, try webbing with the word "Algebra" as your focus. If you are in trigonometry, use the words "Trig Functions" or "Trig Identities." If you are in calculus, use the word "Differentiation," "Integration," or "Series." **Web what you want to stick with you!**

15

Skills to Bridge the Gaps

"Whether you think you can, or think you can't, you're right."

In this chapter, you will find tried and true methods for learning math. These effective study skills are basic to creating a successful, positive experience for yourself.

Six Santa Ana College math students who became math tutors (introduced in Chapter 9)—Enrique De Leon, Jazmin Hurtado, Sarah Kershaw, Joel Sheldon, Alex Solano, and Isabella Vescey—will tell you here what works for them regarding scheduling, class preparation, classroom behavior, note-taking, homework, studying, and reading a math book.

MATH STUDENT/TUTOR ADVICE

BEFORE MATH CLASS

1. Make math a priority.
2. Set aside all the time you need. Expect that math will be time-intensive.
3. Get all of the materials you need at the beginning of the semester.
4. Preview the textbook and your notes before class.
5. Go to math class every day. Be on time.
6. Care for your mind and body. Get sufficient water, food, and sleep.

DURING MATH CLASS

7. Choose your classroom seat carefully.
8. Be prepared by taking your textbook, notebook, paper, graph paper, pencils, erasers, pens, colored pens, calculator, and tape recorder to class.
9. Take questions to class marked in your textbook, notes, or homework.
10. Participate actively in class by taking notes, listening carefully, and asking questions.

AFTER MATH CLASS

11. Take a short break after class, *but* start reviewing and working homework as soon as possible. Do all of the homework.
12. Study in addition to doing homework.
13. Know your teacher, available tutors, and fellow classmates so you have resources when you have questions or need company or motivation.
14. Expect to read your math book with pencil and paper handy to work examples. Write in it.
15. Be persistent. Don't quit. Go to class the whole semester even if you drop the course.

BEFORE MATH CLASS

1. Make Math a Priority

• Before the semester begins, list your goals. Your math goal could be "Pass this class," or it could be more specific such as, "Learn the math in beginning algebra well enough to earn an A or B grade and be skilled enough to succeed in intermediate algebra or geometry." Setting a specific goal will help you set your priorities. Write all your goals for this semester and notice how they fit into your long-range life plan. (If you don't have a long-range life plan, consider making one now. You can change it whenever you wish, but writing one down clarifies your priorities this semester.)

• If you have not already done so, consult a counselor to set up an educational plan that lists your major as well as the courses to accomplish it. Even if you change your major and change the courses that you take, a plan on paper shows that school is a finite process. Knowing that there is an end in sight will assist you in making time for schoolwork and studying.

• Once you have set your long-range goals for your college career and for this semester, set short-term, immediate goals (see Chapter 10) to bring your focus back to living one day at a time throughout the semester.

2. Make Time for Math

Set aside all the time you need. Expect that math will be time-intensive.

• Structure and control your time yourself. Don't let others control it. To build time for math into your schedule, make a sample schedule. Plan your activities for the week ahead by filling out the chart on the next page. Use big categories: Sleep, Travel, Class, Personal Care, Studying, Eating, Socializing, Family Time, Housekeeping Chores, and so on. Build in flexibility to give math the necessary time that it takes.

• At the end of the week, make a second chart with the activities that really occurred. Daily notes in your journal about your activities can help you remember what you really did. Evaluate how you spent your time to discover whether you are using it well. Are you spending *your* time on *your* priorities?

• Plan your time for the next week and track yourself again. Debrief yourself at the end of the second week and continually edit your use of time until it truly fits your priorities. You can increase your pleasure during schoolwork activities by using the skills that you have learned earlier in this book, especially in Chapters 5, 6, 7, 8, 10, and 13. Reread some of these chapters occasionally to remind yourself how you can influence your thoughts and feelings to spend quality time on math.

	Monday	Tuesday	Wednesday	Thursday	Friday	Saturday	Sunday
6 A.M.							
7 A.M.							
8 A.M.							
9 A.M.							
10 A.M.							
11 A.M.							
12 NOON							
1 P.M.							
2 P.M.							
3 P.M.							
4 P.M.							
5 P.M.							
6 P.M.							
7 P.M.							
8 P.M.							
9 P.M.							
10 P.M.							
11 P.M.							

"My advice is to take your time. You won't understand math in five minutes. You won't get it the day before the test. You won't always get it when you're in the classroom. The best thing is to give it time. Just don't stop. If you give it time, you will learn math rather than not learning it and not understanding it and then ignoring it. If you want to learn it in five minutes, good luck. If you want to learn it the day before the test, good luck. What I've noticed is that you've got to give it time. Make time. I don't have a schedule. I'm real loose. I make time. I eat, sleep, and breathe school. So I always have time for school."

—*Enrique*

"I try to make a little schedule like I'm going to try to get this done in this much time. It usually doesn't work out, but I set myself some time goals. Even if I may not meet them exactly, I get some portion of my work done and consider myself successful. That's how I got through last semester. I often had to push things to the next day. Many times I had ideals of finishing ahead and, with all of my classes, I just didn't. But planning my time made me get it done no matter what."

—Joel

"Scheduling is a tough one. My last hurdle is time management preparing for tests. Lots of times I just have to let a lot of other things go. Time and scheduling is the last thing that continues to seem to come up. When I sit down to take the test, I know if I needed to take the time and I didn't take it. One of my best places to do the math is in the kitchen because everything else is there that I need to do—laundry, meal preparation, eating, or answering the phone for my daughter."

—Isabella

"I do better in complete immersion. I find learning little bit by little bit frustrating and confusing. I will sit down with all the materials from the class and in a two-week period devote all of my attention to that class. I do it in blocks. I set aside large blocks of time in which to focus. It has taken a lot of growing up to say 'No' to things that interfere in those blocks of time. If I cannot give it my full attention, I end up making mistakes. And that leads to frustration which leads to a lack of motivation which starts spiraling out of control."

—Sarah

"Success depends on my schedule. It's best for me to take a class every day if I can. It's best to take the midday class Monday through Thursday because I keep it in my head.

"I do what a lot of people call 'marathon studying.' I don't mind studying 10 hours at Barnes & Noble—the whole day on a Saturday or a Sunday. I don't care if it takes me forever to do the homework. I am just excited that I am understanding it."

—Jazmin

3. Prepare in Advance

Get all of the materials you need at the beginning of the semester.

• Buy your math textbook before the semester begins. Of all your textbooks, you will definitely need and use your math book first. Even if you have to borrow a book or the money to buy the book, get it.

• Assemble all of the other materials that you will need ahead of time also. You will need a notebook for math, three-hole notebook paper, graph paper, sharp pencils, erasers, colored pens or pencils, highlighters, and a calculator (scientific or graphing). I like to have a stapler, staples, a three-hole punch, a tape recorder, tapes, and batteries at my desk.

4. Preview

Look at the textbook and your notes before class.

> "Before class, I look ahead at highlighted things in the chapter. I read them. I don't memorize them. I just glance at them, so that when the teacher talks about them in class, they're familiar. Then I say, 'Oh that's what they were talking about.' Everything makes sense. After I preview the material before class and the teacher helps me makes sense of it in class, the homework actually puts it in my head to where I understand it."
>
> *—Enrique*

> "I got into a habit of just briefly reading the highlights of the next section that we were going to cover in class. Just looking at the title gave me an idea. I didn't really look at the examples. I just looked at the bold letters that mean that's something important. I didn't really put much effort into it. I just wanted to see what was going to happen. I didn't want to get caught off guard."
>
> *—Alex*

> "I find it helpful to get a clear-cut guideline of each section—what each section entails and what's going to be in it."
>
> *—Joel*

5. Attend Class

Go to math class every day. Be on time.

> "Show up for class every day and show up on time. If you don't go to class, you miss stuff."
>
> *—Isabella*

> "I have only missed one class in three years of attending Santa Ana College."
>
> *—Alex*

> "I attend religiously." *—Sarah*

> "I go to class every day. I don't do anything else. So if I don't go to class, I don't have anything better to do. What else is there to do during the day? You could watch soap operas. There really isn't anything more important than going to class. You can make some money. What are you going to do with that? It's going to be gone.
>
> "You can't just tell yourself that you don't like math class and go every once in a while. You can't not do your homework and give up. You can't not ask your questions. I think that's death and doom right there. I've seen it happen lots of times. I've seen people disappear. They fade out and disappear."
>
> *—Joel*

6. Care for Your Mind and Body

Your brain is connected to your body. Getting sufficient water, food, and sleep is important. Proper nutrition, and lack of it, affects the chemistry in your mind. Although there is much disagreement about the best "brain foods," there is agreement that insufficient food and water decreases your thinking power. Lack of sleep affects your recall. Pay attention to your health. Food, water, and rest will bolster your math skills.

Bring a bottle of water to class and prepare nutritious snacks or lunches to get you through the day with an alert mind. The time you spend will actually save you time because you will understand more quickly by caring for your body.

"I make sure I eat before class and bring water."   *—Isabella*

DURING MATH CLASS

7. Choose Your Seat Carefully

The choice seat for your learning experience depends on you and your unique learning mode. (Remember Chapter 8.)

"I thought I should be sitting in the back. Maybe because when we lined up at school, I was always told to go to the back. I did like sitting in the back in the corner—not in the middle. Now I can't. I have to be in the front. I need to focus. I need to pay attention. In front, I know I won't get distracted by other things that I see."

—Jazmin

"Sitting in the front works for me but I also found that sitting in the corner is good for me because there's nobody behind me. I experiment. Some days I feel like being in the corner and some days I feel like being in the front. I change my seat and am in control of where I sit and whom I sit next to. I've found that in the past few semesters there are certain people I just can't sit next to. People who are fidgety or have loud, irritating voices throw me off from what I am doing. So I take care of myself. I sit where I'm comfortable with the other students around me and how the lighting is."

—Isabella

8. Be Prepared by Taking Your Supplies to Class

Take your textbook, notebook, paper, pencils, erasers, pens, colored pens, calculator, and tape recorder to class every time. Without your "equipment," you cannot do your job in class. Your choices of materials will be uniquely yours. Give some thought to which ones you really like to use and how you can get them to class with minimal effort. You will not want to be distracted from your learning by a lack of materials. Even sharing a book or calculator can be a distraction.

"I bring plenty of pencils and my calculator."   *—Isabella*

9. Take Questions to Class

Mark them in your textbook, notes, or your homework. Review Chapter 11 on questions.

> "I read my math book with a pencil and I write in the margins. I write questions in the margins that I might ask in class."
>
> —*Isabella*

> "I ask the stupidest question if I need to. I don't let the question go until I get an answer that I understand. It is embarrassing. The people behind start whispering because they get frustrated with me. But that's only one or two people. I have people who come up after class who say, 'Thank you. I didn't get that either.'
>
> "I have found that any question is going to be stupid to some person. It depends on the person. Some people say that any question is okay. We all have that moment. You will be looking at the math problem 7+8 and you can't figure out what it is. Synaptic misfires happen, so tenaciously ask questions."
>
> —*Sarah*

10. Participate

Participate actively in class by taking notes, listening carefully, and asking questions.

> "I always take notes. I copy what's on the board and the examples. I ask questions sometimes to clarify something that I didn't know or understand."
>
> —*Joel*

> "For my notes, I have a separate notebook for each class. I don't mix them. I keep everything all in order."
>
> —*Enrique*

> "Usually I copy what the instructor writes on the board and I write little notes to myself. I also do different colors. I used to do three and I realized that I was spending more time doing notes than listening so I cut down to two. That worked out a lot better. Instead of having to write the same numbers again on that same line I would write a little note in a different color. I actually write everything. I write questions like, 'Why do we do this?' If I am embarrassed to ask in class, I write a little note and try to figure it out by myself or ask somebody. Sometimes the homework answers it. I always write an explanation like, 'We can't do this because of a restriction.'"
>
> —*Alex*

> "Before returning to school, I could never imagine that anyone could take notes in a math class. Now I take notes really fast. I do what they tell you in college success counseling classes. I copy down everything the teacher writes on the board. But I also listen. This semester in statistics, I actually changed things. I listened

more to what the instructor was saying. I listened for his key ideas and wrote them down in words rather than in numbers. Sometimes I would copy off other people's notes—other people that take good notes—in case I missed something."

—Isabella

"Sometimes it's hard for me to take notes in class because I don't want to miss anything or get distracted from the teacher. It's like a story line in a movie. I never want to miss it. I need to get as much information as I can. As the instructor writes, I copy every single little thing, but I have to watch. If the instructor is talking and writing, I listen and copy later. I need to hear him too. At the university, my instructors go really fast so I take a tape recorder. I still try to take notes. If I have key words, I can remember the whole story to it. But if I don't have key words or don't understand something then I'll listen to the tapes."

—Jazmin

AFTER MATH CLASS

11. Do Homework Soon

Take a short break after class, *but* start reviewing and working homework as soon as possible. Do all of the homework.

"I make myself go to the math tutoring center right after my class and not go home. I know that if I go home, I won't do it. It is easier for me to study here on campus. I try to do my homework right after class. The good thing, because I've had class daily, is that I make myself do the homework every day. When I took a class on Monday and Wednesday, I would say I would do my homework on Thursday or Friday and that didn't work."

—Jazmin

"After class I really can't do my homework right away. But I do get to my homework during the same day and I get it done and get it done well. Sometimes I want to start and can't wait—especially in statistics. He gave us a lot of worksheets in class. I would go back and finish those worksheets because basically I had just copied down some things without understanding them. I would not start on the homework but rather on the worksheets so I could go over what we had done in class on that day in the way that we did it.

"I have to practice math more than other subjects. I have to do the homework. My notes from math class are useful to me. I can actually go back to my notes when I can't do the homework and I can see what I'm supposed to do."

—Isabella

"After class would be the ideal time to do math homework, but I tend to feel a little drained after class. I enjoy the math lecture. I think they're interesting, but usually I don't have the time or the motivation to do homework right after class.

I do know that soon after I've learned something is the best time to jump into it. However, it doesn't usually work that way. I do homework anytime I can—anytime before it's due. The most important thing is to understand what all the material is. A good instructor tells you what material you need to know.

"I have to take notes. I have to do my homework and practice problems. Practice does make perfect or pretty good. I identify what kinds of problems there are and know how to do each kind. I recommend knowing the differences between the kinds of problems. I always do all my homework. A homework problem could be on my test."

—Joel

"To do homework, I look at the examples. First I always try to answer my own questions by looking at the book or by looking at the notes. I really don't care how long it takes me to answer. I just want to see if I can answer it myself before I ask anybody else. If I let somebody else answer it, I see what's right but I never put the thought into finding the answer. I think that if I look at the examples I can learn on my own. That helps me remember more."

—Alex

12. Study in Addition to Doing Homework

"I don't think just doing the homework is enough. I do the homework and make sure that I understand the concepts. Once I complete the homework, I ask myself, 'Do I really understand this?' If I don't understand it, I've probably just done the mechanics by following along with an example from my notes or the book. That helps me get the problem but it doesn't help me out in the long run unless I spend time understanding it. When I study, I mostly do homework. I realize that I have been short on time on the studying part. I do the homework part but I kind of missed on the studying part with my calculus class. Studying would have helped me out."

—Alex

"When I study, I recopy the notes. I draw pictures, make arrows, and write things down in steps. I make an introductory sentence about what we're trying to do here as the teacher has said it. I will write, 'We're trying to find the average' and then put the steps to do it. I put a big number one and circle it. I write out the first step in words and then I write out the formula way to do it. Sometimes my numerical illustrations have arrows or captions or bubbles to explain what I am seeing.

"One of my algebra teachers did something that I've carried over to my other classes. He would start the new processes by saying, 'These will be the concepts required.' So I draw myself a little box titled 'Skills Required.' Then I make a list of the skills required such as being able to take an average. That's good for studying. Sometimes to prepare for a test, I'll write down 'Skills Required for Chapter Eight.' Then I'll go through and make a list of all the things I've learned to do. I even have detailed notes with little pictures that I drew of the buttons on my calculator."

—Isabella

13. Meet Your Support Team

Know your teacher, available tutors, and fellow classmates so you have resources when you have questions or need company or motivation.

"I would ask my tutor the questions that I needed to have answered and most of the time everything started to come together. That helped out a lot. My tutor (Enrique) never did the work for me. He would make me explain it to him and I think that was the key part to it. If I couldn't explain it, that meant that I didn't understand it. If I explained it, I would have confidence in myself, knowing that I could explain it to somebody else. I talked more than he did. He would just listen and correct me if I was wrong. He would say, 'That's correct' or 'You want to do this first.' If he did the work, I would always say, 'Oh, I understand it.' It would be like going to class when the instructor does the problems. I always understand them in class but when I sit down to do the homework, it's like, 'Oh no.' Similar problems but I can't do them. There's just something that's different when somebody else does the problem for you and when you do it yourself.

"I'm glad I've made good friends with the tutors. They're one of my resources. I get along with the math department and all my instructors so I feel comfortable talking to them. I get comfortable with someone when I see them a lot. I remember when I went to my instructor's office the first time to ask a few questions, I was kind of embarrassed but I said, 'I have to do this or I won't understand.' Probably for someone to be comfortable with an instructor they need to go to their office hours. In class you can only talk to them for a little time before class or during class or after class. But in office hours, they're there. I go to office hours. The instructors are sitting down and they're there to listen to you."

—Alex

"Sometimes when I sit down to study for a test, I really don't know where to begin at all. So this last semester I joined a study group. That really helped because I spent hours and hours with them, which was hours and hours that I wouldn't have spent at home. Plus I made an effort to prepare before we met so I'd be ready to do something when we got together. We started with Chapter 1 and studied with the instructor's worksheets. We all got copies of our tests and helped each other with the ones we missed. When I took the final, those were the things that I did the best on. When people talked about the final afterwards and what they had forgotten, I got it because we had studied it. There were nine chapters on the department final. In the beginning we did the first three chapters because they were short. Then each chapter got bigger. There were just three of us and sometimes a fourth person who would float in and out or a different person who showed up and never came back. We reserved the room in the back of the library to use the white board. We met at the table and hashed over stuff, staying an hour or two, two or three times a week. A group of students met on Fridays from 9 A.M. to 12 noon. They invited me and I started going to that for an hour too. We'd just add another day. By the last week we were meeting about every day during the week."

—Isabella

14. Read with a Pencil

Expect to read your math book with pencil and paper handy to work examples. Write in it.

"I read from the beginning to the end of a section. I'm not memorizing. There're billions of examples. Everything the book says is backed up by an example. As I read, I grab paper and pen and write down the example. I always have paper and pen at hand. I read and then I write. I've tried just reading it and I can't. My hand starts twitching and I say, 'Give me that pen and paper. I've got to see how they do it.' Wow! It makes a really big difference.

"Every time I tutor someone in math, I ask them, 'Have you read your book? Have you read that chapter? Have you read that section?' They always say no, as if a math book can't be read. I tell them they could read their math book. They just don't know how. I tell them how I read it. I read and, when it says, 'Example,' I do the example. That really helps. I don't want to memorize everything. By doing that example I remember the definition and know what it's about. Sometimes I hit a roadblock and I go back and do the example again. I carry examples with me to review. I never know when an example will help me on a test."

—Enrique

"I started reading my math book when I repeated my algebra class, and it helped out a lot. I realized that the wording is very difficult and I just kept practicing. After a while, it seemed like my language. I didn't have to say 'what does this mean?' I started reading it just as if it were a regular book. I was surprised. It's not easy but it started to work for me. Helping other students out with their math, I used the same language over and over, so reading the book became much easier. I do work examples through with paper and pencil. I write down the question. Even if it's a word problem, I write it down. Because when I write it down in my own handwriting, I understand it more as compared to just reading it. I always work out the problems. I write little notes in my book. I will write little marks like question marks. I don't do my problems in it. But I write little notes—anything that will help me out in reading the book. If I see a sentence with a vocabulary word, I'll write the definition to it instead of writing it on a separate paper and looking back and forth from the paper to the book. I just write it in the book so it's all in front of me."

—Alex

"Most of the time, I have to read the book. Usually I go to lecture and then read the pages related to the lecture in the book. At the same time I read the chapter, I look at my class notes so I kind of do the whole lecture again in my head while I'm reading it. I put those two things together at the same time. That helps me to understand the chapter more and the notes better."

—Jazmin

"Whether or not I read the math book depends on the book. Since I learn best from the whole to the specific, if it explains the theory, I actually read it. If the math book talks about rules or simply the details of how to do something, I don't because that confuses me. I can't keep all of the rules straight. Because of dyslexia, they start dancing the jig.

"So I will look back at examples if I encounter a problem and have that moment of brain freeze and cannot for the life of me figure it out. But, again, if the book or examples do not explain the theory, I will find a way to fit it into my preexisting constructs."

—Sarah

"I don't always read the book. I try to get by with as little reading as I have to. It takes a long time to do the homework and sometimes I could sit there and read the book and not understand what they're talking about. Sometimes I read the text ahead and then jump into working some of the problems. Then I think, 'Oh, O.K., that's what they're talking about.' Other times I just jump into doing homework and then I read back in the book if I need to."

—Joel

"I write in my book with a pencil. I break examples down. I make little arrows. I read the chapter but I seldom get to finish. I usually get at least half or three quarters read.

"I do what you are 'supposed to do' with all college material. I read the summary first and then go back and read the chapter. And now I review the entire math book at the beginning of the semester. They taught us this in study skills. It's called 'Textbook Reconnaissance.' You open the book and begin with reading everything on the covers, skim every page at the beginning, read the table of contents, go through the preface and introduction. That gives me a lot of information about what's going to be done during the semester. You learn if there are Web sites, which I used a lot one semester because the Web sites had quizzes. If the book has an introduction to calculators, I would look at that. For each chapter you don't read everything. Look at headings. Look at drawings. Look at sidebars with applications. Look at everything in bold print. It's a nice weekend activity if no one's around for a couple of hours. You can spend time with your new textbook. The algebra book and statistics book had sections on how people use math in their jobs. I remember one on counting moose."

—Isabella

15. Be Persistent

Don't quit. Go to class the whole semester, even if you drop the course.

"The first time I took Calculus II, a month before the end, I explained to the instructor that I was going to drop it. He agreed that was the best idea, but I asked him if I could please attend the lectures because I was going to take it again that summer. I knew going through it once already was going to help me. It would make it easier the second time.

"When he explained things in class and I didn't understand, I copied the notes down. I wrote all of the details, even things he would say. I knew that I was going to get it. Maybe not that day, maybe the next day or whenever, but I knew I was going to get it. I got excellent notes from that instructor. The best. When I took it again that summer, I used my first instructor's notes. I took them with me to the class the second semester with the new instructor. I didn't understand the new instructor and I couldn't get his notes or his style. I couldn't get anything. A lot of times I just had my first instructor's notes in front of me when the second instructor was lecturing. When I couldn't understand him, I would look over the first instructor's notes. They were good notes. Very good. And I passed. I got it."

—Jazmin

"I just walked through the feelings and the panic. When I would get tests back that were C or D grades, I would just say, 'Oh, well. It's just a test.' I knew I would be coming back the next class meeting. For me, there wasn't any option.

"My advice is to show up for class every day and show up on time. If you don't go to class, you miss stuff. If you're sitting there in class in an effort to pay attention—even if your mind is wandering—you're getting something. At least try to do the homework—at least try. If you get stuck, go on to the next problem.

"Keep taking classes over again. Just keep taking them until you are ready to transfer or ready to graduate."

—Isabella

"My worst math experience was getting a D in algebra. I thought it was the end of the world. It was very hard for me. I didn't know what I was doing wrong and what I was doing right. I got off to a good start with a B. Then everything went by so fast. I felt awful. I was O.K. with prealgebra and beginning algebra but, in second semester algebra, I was always struggling the whole semester. I thought math was impossible for me. It lowered my self-esteem. I thought, why couldn't I do math? I saw everybody getting hundreds. When I would go to the Math Study Center, I would hear people use big mathematical terms. **But now I'm just as good as they are at that.**

"My advice to struggling math students is 'Do not give up.' I thought I was going to give up because math had never been my favorite subject. I never gave up on it even though I struggled a lot."

—Alex

Pushing Your Limits

CHAPTER

1. With which math student/tutor do you identify? Why? What was your reaction to what they said? What is your favorite quotation? Write it in your journal.

2. Reread the 15 points in this chapter. Check those that you want to remember. Xerox all 15 and tape them in the front of your math book or write them in your journal. Which of the 15 points are you willing to do now that you haven't done before?

3. Reread point #1 and do the suggested goal-setting exercises in your journal.

4. Make an appointment with a counselor to develop your educational plan as suggested in point #1. Remember, the plan can always be changed, but writing it down will help you make it happen.

5. Reread point #2 and do the scheduling exercises in your journal. Leave blank space to return and continue the process of taking control of your time.

6. Working at spatial visualization skills can improve your math skills. Mastering Math's Mysteries, Chapter 15, will show you one way to improve.

Spatial Visualization

You don't have to be good at spatial visualization to be good at math, but it helps. You can improve your spatial visualization skills by working puzzles, sewing, reading maps, making maps, drawing, building, and using math manipulatives such as blocks, counters, or Tangrams. Sketching on Geopaper (see the Appendix) or graph paper also improves your spatial visualization.

As you work through the exercises of this worksheet, play. Let the pressure and the competition go. If a puzzle seems impossible, take a break and return to it later. There are many ways to improve spatial visualization. It is a slow, steady process.

Tangram Puzzle

The Tangram Puzzle is an ancient Chinese puzzle with seven pieces—five triangles, one square, and one parallelogram. You can make thousands of interesting shapes with Tangrams. Trace the Tangram pieces that follow onto cardboard, buy a set of Tangrams at an educational or office supply store, or borrow a set from a math teacher. Then do the activities.

Tangram Activities

1. Make and draw a shape of your own using all seven pieces of the Tangram puzzle.

2. Fit the seven Tangram shapes onto the following puzzle.

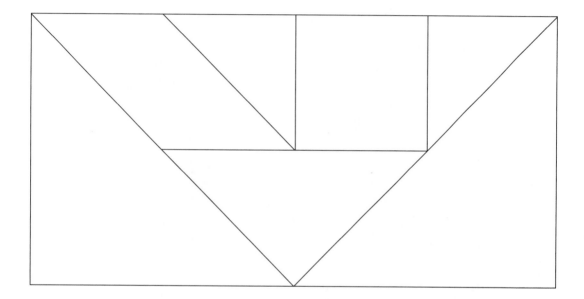

3. Compare the sizes of the triangles. (a) Put the two small triangles together to make a medium triangle. (b) Use all three of the small and medium triangles to make one large triangle. Sketch your results.

4. (a) Use the two small triangles to make the parallelogram. Sketch it. (b) Then make the square and sketch it.

5. Compare how the two small triangles make the medium triangle, the parallelogram, and the square. Describe the comparison in writing.

6. Put your Tangrams together to make puzzles a, b, and c on the next pages. Do not be dismayed if you cannot do them immediately. Leave and come back. Have the neighbor kids help you. **Make it a game!**

7. For a really big challenge, make a square using all seven pieces.

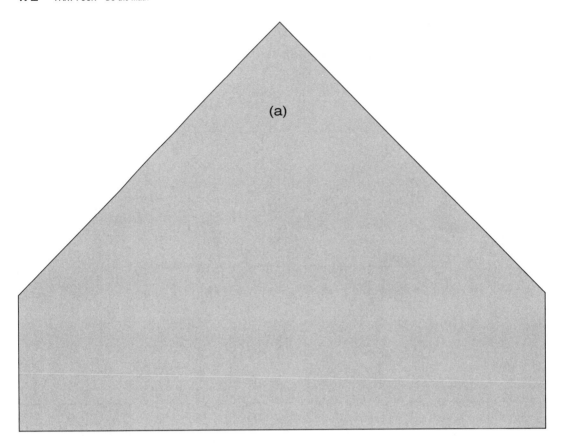

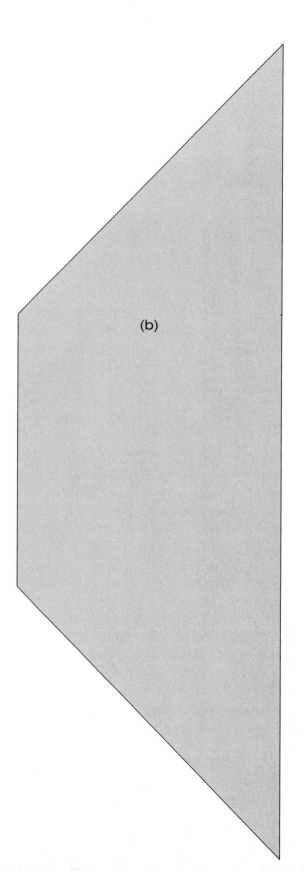

(b)

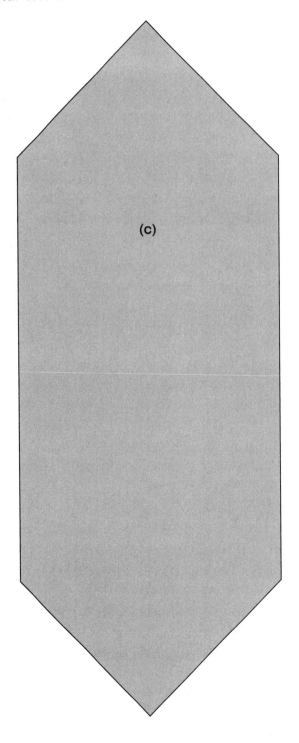

16

Creative Problem Solving

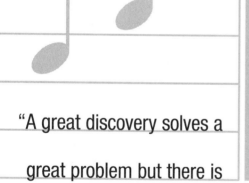

"A great discovery solves a great problem but there is a grain of discovery in the solution of any problem."

G. POLYA

"We've racked our brains over this one for quite some time now, and, as I say, the cleverest number devils have tried every trick in the book. Sometimes we can work it out and sometimes we can't."

HANS MAGNUS ENZENBERGER

POLYA'S FOUR STEPS

There is no "one best way" to solve problems. There are simply many useful methods that could work. Fill your "bag of tricks" with strategies so that they are available to you. Add to your repertoire with experience and the willingness to take risks.

Polya's Four Steps

1. **Understand the Problem.**

2. **Devise a Plan.**

3. **Carry out the Plan.**

4. **Look Back.**

In 1945, Stanford mathematics professor George Polya wrote a little book entitled *How to Solve It*. It has remained a best seller because it relates understandable methods to solving problems in many fields.

Polya broke problem solving into four amazingly simple, but powerful, steps:

1. Understand the Problem.
2. Devise a Plan.
3. Carry out the Plan.
4. Look Back.

Polya's steps can be applied to any problem, whether it is a real-life problem, arithmetic problem, algebra problem, geometry proof, word problem, calculus problem, or virtually any other kind.

The first and last steps are commonly ignored. Most students spend too little time trying to understand the problem in the beginning. They want to immediately jump into working it and getting an answer. Once they have the answer, they want to be finished and not look back. They hurry on to the next problem, only to be stumped by similar problems down the road because they did not internalize the process.

In this chapter we will focus on problems that illustrate Polya's procedures. Reread his words at the beginning of the chapter. If you are willing to look for the "grain of discovery" in each of these problems, your "bag of tricks" will get bigger and bigger. Then as you attempt new problems, a wealth of strategies will come to mind that you already understand and can relate to the problem's circumstances.

As you proceed through this chapter, the methods, not the problems themselves, are important. How they are solved should be your focus. For further practice, look into excellent books such as *The Only Math Book You'll Ever Need*, by Stanley Kogelman and Barbara R. Heller (1994).

Wait a Minute!

Hold everything! Before jumping into Polya's Steps, let's talk about what to do if you get stuck. Here are some time-honored **mind-shifting practices.** Try them to stimulate shifts in your mental processing.

1. **Choose your work location carefully.** You will want to give your whole attention to this study. Find a place that suits your unique needs for space, sound, and sight. Give yourself the privacy that you need. Switch sites periodically. A headset with a CD can make a public place into a private study sanctuary.

2. **Break often.** Breaks can be short and as simple as closing your eyes, taking a deep breath, stretching your arms, or moving your head gently to stretch your neck. Periodically, walk. Get a drink or duck outside for fresh air. Especially when you are stuck, stop a while and release your thoughts.

3. **Ask questions.** Then bring yourself back to your work and continue.

4. **Review.** Perhaps some previous idea was not clear enough or you misunderstood. Begin the chapter or the section again, working your way through by reading and doing problems. Say to yourself, "I must have missed something." In math, every word and concept is important. Ignoring or misunderstanding one key word could throw you off completely.

5. **Use your voice.** Reading the problem aloud puts the details into your mind auditorily and visually, whereas reading the problem silently only puts it into your mind visually. There is a difference. The more parts of your brain that you involve, the more you will understand, increasing the likelihood of reaching a solution.

6. **Read backward.** Reading the problem backward, word by word, forces you to focus on the problem, slowly breaking up your preconceived ideas and bringing individual words to consciousness.

   backward. problems word Read

7. **Hang in.** Don't give up. Persevere. Eventually it will come. You may have to wait a while, and it is O.K. to call in assistance. Do whatever needs to be done.

8. **Let it go for the moment.** Dropping a specific problem and working others releases that problem so that your subconscious can take over and do its work.

MIND SHIFTING AT WORK

Mind shifting allows energy to be released so that your subconscious can solve the problem. As you experiment, you will discover what works for you. Here are two people who became famous for the problems that they solved, and stories about how the "light bulb finally turned on."

ONE BUS RIDE. Famed botanist **Charles Darwin** spent seven years traveling the globe studying plant and animal life. On his return to England, he worked long hours for years classifying his specimens. Connections he believed to exist eluded him. One day as he *rode in the top of a two-story bus* across London, his work suddenly came together, forming basic ideas for his theory of evolution (Stone, 1980). With that one idea, he revolutionized man's understanding of life on earth.

FLY WATCH. My favorite math story is about **Rene Descartes** lying on his bed as a sickly teenager watching a fly walk the ceiling. In a flash, he recognized that the fly's position could be described with two numbers—the perpendicular distances from the fly to two edges of the ceiling. With this basic idea, he eventually developed the rectangular coordinate system for graphing relationships. *Not while concentrating but while daydreaming*, Descartes discovered the fundamental concept that allowed algebraic symbols to be visualized geometrically. He became one of the world's most influential scientists.

Although you and I may or may not discover ideas or solve problems of this magnitude, we can learn from the creative process of Darwin and Descartes.

- They studied intently, absorbing information and ideas into their minds.
- They worked with their ideas.
- They were curious with direction and intention.
- They were tenacious, yet they took breaks.

It's time for you to practice some problem-solving techniques yourself. Let's go back to Polya's Four Steps and practice strategies. If you work hard here, you will find your overall anxiety level decreasing in math. The more word problems that you successfully solve, the more confident you will be approaching new problems. Experience is a great teacher.

REVISITING THE FOUR STEPS

Step 1. Understand the Problem

What are you asked to find? What is the situation? What are the conditions? What do you know for sure? What do the words in the problems mean? Begin writing, sketching, organizing, or modeling this information. Here are three strategies (Afflack, 1982) to increase your understanding and a problem to illustrate them.

STATE THE GOAL OF THE PROBLEM SPECIFICALLY. Define what you seek. Get a focus. Write down what you are asked to find. How will you measure or recognize it? Could you have misunderstood? Often the goal or request is stated at either the beginning or the ending of a problem. Write "Find . . ." and complete the sentence, such as "Find the speed in mph."

DECIDE WHAT IS KNOWN AND UNKNOWN. What does the problem tell you specifically? What don't you know yet?

THROW OUT IRRELEVANT INFORMATION. Which information has nothing to do with what you are asked to find? Eliminate it and use the rest. The problem may give details, such as the color of the car or the gender of the people, that have nothing to do with what you are requested to find.

Example 1: Selling Tickets

Andreas and Joe need to report on play tickets they sold in their neighborhood of 35 houses. The tickets cost $5 for adults and $2 for children. So far, they know they've sold 23 tickets and collected $76, but they must make a report of how many adults and how many children will attend. Can you help?

What is the goal of this problem?

Andreas and Joe must report how many adults and how many children will attend. Writing this specifically focuses our search. "Find the number of adult tickets sold and find the number of child tickets sold."

What information is known?

There are 35 houses in their neighborhood.

The tickets cost $5 for adults and $2 for children.

They sold 23 tickets. (Total tickets.)

They collected $76. (Total money.)

What information is unknown?

The number of adult tickets sold.

The number of children's tickets sold.

What information is irrelevant?

The number of houses in the neighborhood has nothing to do with tickets.

Example 1 will be continued under Step 2.

Step 2. Devise a Plan

Have you seen a problem like this before? Do you know any relationships between what you are asked to find and the conditions you are given? Are there any formulas that relate? Is there a way to restate the problem more simply? What strategies might fit? Here are nine strategies (Afflack, 1982) for Step 2, with examples. Work through each example with pencil and paper.

TRY SOMETHING "OFF THE WALL." Brainstorm. Draw outside the lines. Hunches or intuitions or guesses may seem risky, *but* they can break through to a solution if you work with them. Mathematics was created from intuitive guesses and dreaming. The logic binding it all together came later—sometimes hundreds of years later. Think differently from a new perspective.

Example 2: Karl Friedrich Gauss

Famous mathematician Karl Friedrich Gauss (1777–1855) was 10 years old when he and his unruly classmates were assigned work to keep them out of mischief. They were told to add the counting numbers from 1 to 100. Imagine the teacher's amazement when Gauss came up with the correct answer, 5050, in a matter of minutes. Asked to explain, Gauss said he paired 1+99, 2+98, 3+97, . . . , 48+52, and 49+51, which was 49 pairs each adding to 100. Adding 49 • 100 plus the numbers 50 and 100, which weren't paired with anything, he came to the total answer, 5,050.

Example 3: The Boys' Room

I hurried into the women's restroom at Dodger Stadium during an exciting baseball game. The restroom had two sides of stalls with a bank of sinks in the center. Just inside the door, a small boy wailed to his unhappy mother, "I want to go to the boys' room." The young mother looked frustrated and helpless standing over the sobbing child, who was too young to visit the men's restroom alone. While they interacted, I noticed an available stall in the corner. No women were washing their hands nearby. A thought popped into my mind. I leaned over, pointing to the empty stall, and said to the boy, "There's the boys' room." He stopped crying and ran to the open door, followed by his clearly relieved mother. My "off the wall" white lie temporarily converted the gender of one Dodger facility, but it solved a rather loud and urgent problem.

GUESS AND CHECK. Guess and test your guess. Make an easy first estimate and see if it works in the problem. Record your guess. Of course, it is wrong. It is only a guess. All that you care about is whether the first guess is too large or too small. Refine your second guess accordingly. Continue refining each guess, testing it to see how far off you are. You might be surprised how quickly you can solve your problem.

Example 1 Continued: Selling Tickets

Reread Example 1. We want to find how many adult tickets and how many child tickets were sold.

(a) If we **guess** that 5 adult tickets were sold, Andreas and Joe must have sold 18 child tickets. Why? Because they sold 23 tickets all together. Let's **check.** Five adult tickets cost five times $5 each, for a total of $25 for adult tickets. How much would 18 child tickets cost? Eighteen child tickets would cost eighteen times $2 each for a total of $36 for child tickets. Adding the two total costs, $25 for adults and $36 for children, means that our **guess total** is $61. That is **not enough** money, since they collected $76.

(b) Somehow we **need to get more money.** Since adult tickets cost more, let's **guess more** adult tickets. If we guess that 8 adult tickets were sold, Andreas and Joe would have sold 15 child tickets. (Remember that they sold 23 tickets all together.) To **check,** eight adult tickets cost 8 ($5 = $40. Fifteen child tickets cost 15 • $2 = $30. Total money collected for this guess is $70. We are **getting closer** to the money they actually collected ($76).

(c) We still **need more** money so let's **raise our guess** for adult tickets to 10 adult tickets. That means Andreas and Joe sold 13 child tickets (23 total tickets sold). **To check,** ten adult tickets cost 10 • $5 each = $50. Thirteen child tickets cost 13 • $2 each = $26. Totaling the cost: $50 for adult tickets + $26 for child tickets = $76, the amount of money collected by Andreas and Joe. It works!

Our third guess gives the correct answer: Andreas and Joe sold 10 adult tickets and 13 child tickets. (This answer gives 23 total tickets costing $76, which checks with the original problem.)

This table documents our three guesses and their checks.

	# adult tickets	# child tickets	$ adult tickets	$ child tickets	Total $ collected is $76
(a)	5	18	5 • $5 = $25	18 • $2 = $36	Total = $61 Not enough
(b)	8	15	8 • $5 = $40	15 • $2 = $30	Total = $70 Closer, not enough
(c)	10	13	10 • $5 = $50	13 • $2 = $26	Total = $76 Correct!

SIMPLIFY THE PROBLEM. If possible, bring the problem level down to one case or a few cases. Solve the simpler problem and then work up one step at a time, discovering patterns. Suppose that you are asked to find how many trips 50 people make. First, think about one person. Here's another example.

Example 4: Family Reunion

The Thomas family (eight people) is about to have a reunion. A very friendly family, they like to stay close one-on-one by telephone. If every family member needs to speak with every other family member once, how many calls does that make?

We are expected to find the total number of calls that eight people would make if each one spoke with every other person. Let's think about smaller families first. In fact, let's start with a family with one person. That's zero phone calls. (A tiny phone bill.) Next, let's think about a family of two people. In a family of two people, there would be one phone call. Moving up **slowly,** let's consider a family of three people. The first person calls the other two. That leaves only one phone call for the other two to talk. A total of three phone calls for three people. Here is that information organized in a t-table. We can find more information by using diagrams to figure the number of calls for families of four and five.

# people	# calls
1	0
2	1
3	3
4	6
5	10

Each dot is a person.
Each line is a call.

Notice that each time we add another person, the difference in the number of calls increases by one. (See below.) You might also notice that the number of calls is the sequence of Triangular Numbers—a common sequence of numbers that we looked at in Chapter 5's Mastering Math's Mysteries. Using this pattern, we discover that an eight-member family makes 28 phone calls.

# people	# calls
1	0
2	0 + 1 = 1
3	0 + 1 + 2 = 3
4	0 + 1 + 2 + 3 = 6
5	0 + 1 + 2 + 3 + 4 = 10
6	0 + 1 + 2 + 3 + 4 + 5 = 15
7	0 + 1 + 2 + 3 + 4 + 5 + 6 = 21
8	0 + 1 + 2 + 3 + 4 + 5 + 6 + 7 = 28

ACT IT OUT OR USE OBJECTS. Model the problem. For new perspectives, move objects or yourself around to develop a visual, kinesthetic image of the situation. Use a little drama. If a problem has motion in it, simulate the motion so that you clearly see what happens. Einstein reportedly came up with his famous formula $e = mc^2$ by visualizing himself running alongside a beam of light. Sometimes he pictured himself riding the beam. Einstein experimented with this visualization for years before breaking through to understanding. Are you willing to be that persistent?

Example 5: The Garcias

There are seven Garcia kids—Sonia, Jose, Juan, Elena, Carlos, Luisa, and Henry. Luisa is older than Jose and Carlos. She looks up to her big sister Sonia, who is not the oldest child. Juan is older than Carlos but not Jose. Carlos is not the baby, and Henry says that he's tired of babysitting his siblings. If the Garcia family has no twins, list the kids in order from the oldest to the youngest.

The goal here is a list of children from the oldest to the youngest. Use **seven small pieces of paper** with names—each representing a child. Write each name on a slip of paper and set them aside. Sentence 1 gives you their names.

Sentence 2 puts Luisa above Jose and Carlos, but we don't know whether Jose or Carlos is older, *yet.* Place the paper with Luisa's name on it above in your work space, with Jose and Carlos side by side below for now. Keep the other names out of your work space until you have a place for them.

Sentence 3 puts Sonia above Luisa, since Sonia is Luisa's big sister. Sonia is not the oldest, so there is an empty place above Sonia that we can't fill *yet.* Place Sonia's paper at the top. Use a blank piece of paper or any object to hold the empty spot above Sonia for the oldest.

Sentence 4 puts Juan between Carlos and Jose, with Jose as the oldest of those three. *So far,* your pieces of paper from the top down will be: Empty Place, Sonia, Luisa, Jose, Juan, and Carlos. Which pieces of paper are left? Elena and Henry. That means that one of them must be the oldest, but we don't *yet* know which. Read on.

Sentence 5 puts Carlos above someone, since he is not the baby. Now, there is an empty spot at the top and the bottom. Since this sentence also says Henry is tired of babysitting, he must be the oldest.

The siblings are Henry, Sonia, Luisa, Jose, Juan, Carlos, and Elena from oldest to youngest.

MAKE A PICTURE OR A DIAGRAM. Use pencil and scratch paper to draw what is happening. Find a way to sketch the relationships. Organize the details graphically on paper. The results do not have to be artistic—you are merely making a visual simulation of the problem.

Example 6: Driving

Don will be driving north at 75 miles per hour while Luisa drives south at 60 miles per hour. How long will it take them to be 405 miles apart?

We need to find the amount of time it takes for them to be 405 miles apart. Let's draw what's happening. They are going in opposite directions. After just one hour, Don has driven 75 miles and Luisa has driven 60 miles. (That's what 75 mph and 60 mph mean.) Each hour, Don drives 75 miles more and Luisa drives 60 miles more. The three diagrams below show what happens each hour. After one hour, they are 135 miles apart. After two hours, they are 270 miles apart. After three hours, they are 405 miles apart—our goal. The diagram below shows how it takes three hours for them to be 405 miles apart.

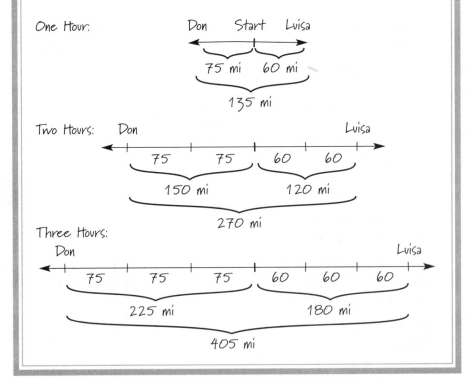

Example 7: The Diaz House

The Diaz family just moved into a brand-new housing development. They are thrilled with their new home and want it to look like the decorator model home, where Mrs. Diaz noticed that the crown molding around the edges of the master bedroom ceiling was quite attractive. On the sales brochure, she sees that her master bedroom is 11 feet by 13 feet. Her budget for this project is $200. If the materials and labor for crown molding cost $10.00 per yard, can she afford this upgrade?

To see if Mrs. Diaz can afford the molding, we need to know her cost and compare that to the budget of $200. To find her cost, we need to know how many yards the job requires. The room measurements are in feet, so a drawing of the room will be a start. See the room diagram below with the work.

```
        13 ft.                  13          3 feet = 1 yard
                                11          48 feet = 16 yards
                                13
11 ft.          11 ft.        + 11
                              48 feet around

        13 ft.                Cost is 16 yards • $10 = $160.
```

Since the crown molding cost is $160, which is less than her $200 budget, Mrs. Diaz can afford it.

ORGANIZE THE DATA. Using a t-table or a chart, list the information in an orderly manner as you discover it. Arranging the data systematically shows you what you know and don't know. Tables help you see any developing patterns and assure that you have considered all of the possible cases.

Example 8: Malena's Money

Malena found six coins in her pocket. If she has at least one quarter, at least one dime, and at least one nickel, how much money could she have?

(continued)

Example 8: Continued

We are asked to find the possible values of Malena's money. This problem has more than one answer.

First, let's concentrate on the possible number of quarters, filling in with the numbers of dimes and nickels. If Malena has six quarters, she needs no dimes and no nickels, which does not work since she must have one of each type of coin. If she has five quarters, she needs only one other coin and could not have both a dime and a nickel.

See how the table below systematically changes the possibilities by slowly decreasing the number of quarters and then the number of dimes. Some of the combinations cannot occur, since she needs at least one of each coin.

# QUARTERS	# DIMES	# NICKELS	
6	0	0	No dimes or nickels
5	1	0	No nickels
5	0	1	No dimes
4	2	0	No nickels
4	1	1	$1.00 + $.10 + $.05 = $1.15
4	0	2	No dimes
3	3	0	No nickels
3	2	1	$.75 + $.20 + $.05 = $1.00
3	1	2	$.75 + $.10 + $.10 = $.95
3	0	3	No dimes
2	4	0	No nickels
2	3	1	$.50 + $.30 + $.05 = $.85
2	2	2	$.50 + $.20 + $.10 = $.80
2	1	3	$.50 + $.10 + $.15 = $.75
2	0	4	No dimes
1	5	0	No nickels
1	4	1	$.25 + $.40 + $.05 = $.70
1	3	2	$.25 + $.30 + $.10 = $.65
1	2	3	$.25 + $.20 + $.15 = $.60
1	1	4	$.25 + $.10 + $.20 = $.55
1	0	5	No dimes
0			No quarters

There are 10 answers. Malena could have $1.15, $1.00, $.95, $.90, $.80, $.75, $.70, $.65, $.60, and $.55. The table covers all of the possibilities assuming she has no pennies or half dollars.

IDENTIFY PATTERNS. When the data is organized, patterns are easier to identify. Notice any pattern—relevant to the problem or not—but do not get invested in keeping a particular pattern. Let go of anything that doesn't fit. A t-table or chart that illuminates patterns helps you to get more data or to generalize the relationships into a formula.

Example 9: Mowing the Lawn

Amy mowed her rich old uncle's lawn. When she had finished, he said, "I will either pay you $100 today or I will pay you 1 cent today and double it each day for a whole month. Which do you prefer?" Now, Amy had no pocket money at the moment, and she very much wanted to go to the movies with her friends who were standing nearby. They were absolutely dumbfounded when Amy said loud and clear, "I'll take the penny." Why would she do that?

Amy must have thought fast. Maybe she used her fingers to keep track of the days, doubling the money as she moved across her hands. Try it. One finger = 1 penny. Two fingers = 2 pennies. Three fingers = 4 pennies. Four fingers = 8 pennies. Five fingers = 16 pennies. Six fingers = 32 pennies. Seven fingers = 64 pennies. Eight fingers = 128 pennies. On the eighth day she finally gets more than a dollar. Nine fingers = 256 pennies. Ten fingers = 512 pennies or $5.12.

Let's round $5.12 down to $5 and keep going with the fingers. Eleven fingers = $10. Twelve fingers = $20. Thirteen is $40. Fourteen is $80. Fifteen is $160. Wow! If Amy can wait until Day 15, she receives over $100 in just one day. And, after Day 15, the money keeps on doubling!

Amy gets very curious about how much money she'll receive on Day 30 so she says goodbye to her friends and walks home for a pencil, paper, and calculator. Recording each day on a t-table and beginning again, Amy notices that the amounts of money are always powers of two. That means that each amount of money can be calculated by multiplying twos together. Look at Amy's t-table. Day 2 is 2 cents. Day 3 is 2 • 2 or 4 cents. Day 4 is 2 • 2 • 2 or 8 cents. And so on. Amy notices a pattern. She sees that the number of twos is just one less than the number of the day. She continues the t-table forward to check out this pattern. Yes, Day 5 is

(continued)

Example 9: Continued

four twos multiplied by each other to get $.16. Day 6 is five twos multiplied by each other to get $.32. Etc.

Day Number	$
1	1 penny
2	$2 = 2 = 2^1$ ←
3	$4 = 2 \bullet 2 = 2^2$ ←
4	$8 = 2 \bullet 2 \bullet 2 = 2^3$ ←
5	$16 = 2 \bullet 2 \bullet 2 \bullet 2 = 2^4$ ←
6	$32 = 2 \bullet 2 \bullet 2 \bullet 2 = 2^5$ ←
7	$64 =$
8	____
9	____
.	
.	
.	
30	____

Amy sees that this exponent is one less than the "Day Number."

   Since Amy knows about exponents and her calculator, she knows that $2 \bullet 2 \bullet 2 \bullet 2$ is written 2^4 and she can use the x^y calculator key to verify this. (Calculators differ, so check your manual.) Amy checks her calculator technique by checking that $2^5 = 32$ and $2^6 = 64$ before she even tries to calculate 2^{29}, the amount of money that her uncle would give her on Day 30. When she is sure she knows how to calculate powers of two, Amy takes a deep breath and punches in 2^{29}. She takes a look at the answer and screams, "I'm rich." How much money will she receive on Day 30? Would you have made the same choice as Amy? Can Amy's uncle afford it? What happens on Day 30?

WORK BACKWARD. Starting at the end of a sequence of events and working back in time cracks some problems wide open. This strategy also presents the data from a different perspective and opens your mind to new thoughts about the problem.

Example 10: Baking Cookies

Emilio baked cookies for his family and friends. He burned the first dozen cookies and threw them out. Then he put away half of what remained for his younger brothers. Half of what was left he gave to his friend Julia. Afterwards he ate half of what remained. If Emilio ended with six cookies, how many did he bake?

Find the total number of cookies that Emilio baked. Begin at the end, when Emilio has six cookies, which are the half that Emilio did not eat. Emilio must have eaten six cookies. One half is the six remaining, and the other half is the six he ate. Those are the 12 cookies that he did not give Julia. In fact, they are the half she didn't get, so she got 12 cookies. One half is Julia's 12 cookies and the other half are Emilio's six and the remaining six. Look at this diagram to see what is happening.

6 left.	Emilio ate 6.
Julia gets 12.	
Put away 24 for younger brothers.	

plus one dozen burned cookies.

```
    6  left
+   6  for Emilio
+  12  for Julia
+  24  for brothers
+  12  burned
   60  cookies
```

Continuing back to the beginning, Julia's 12, Emilio's six, and the remaining six cookies (or 24 total) are the half that were not put away for the younger brothers. That means the brothers get 24 cookies. This time, one half is the 24 for the brothers and the other half is eventually shared (unevenly) by Julia, Emilio, and the leftover plate.

Since Emilio burned 12 cookies, the total number of cookies that he baked is 60.

USE ALGEBRA. Representing the unknown (what you don't know) with a letter and manipulating symbols solves a wealth of problems. We will rework Example 6 to see how to use algebra.

Example 6: Reworked

Reread Example 6. Remember that we know the answer is three hours. This time we will find that answer by representing what we are seeking by a letter. We will practice Polya's Four Steps here too.

Step 1. Understand the Problem. We know that Don and Luisa drive opposite directions. Their speeds tell us how far they will travel each hour. Don will travel 75 miles every hour. Luisa will travel 60 miles every hour. We want to find the number of hours it will take them to be 405 miles apart. So: Find the time it takes them to be 405 miles apart.

Step 2. Devise a Plan. We plan to use algebra. Let's represent the number of hours with the letter t. Using the diagram that we drew previously for Example 6, we see that our goal is:

Don's distance + Luisa's distance = 405 miles.

This is called an equation (Left Side = Right Side). Also notice from our previous diagram that Don's distance is the number of hours times 75 and that Luisa's distance is the number of hours times 60. Since t represents the number of hours,

Don's distance = $t \cdot 75$ and Luisa's distance = $t \cdot 60$

so the equation from above becomes:

$t \cdot 75 + t \cdot 60 = 405$

Step 3. Carry out the Plan. We have a plan, and it is to solve an equation to figure out what the letter t could be. Don't worry about solving; I will walk you through it as best I can. You will learn how to do this when you take algebra. In solving an equation, the left side always has to balance the right side.

$t \cdot 75 + t \cdot 60 = 405$

When you multiply, order doesn't matter. Notice that $2 \cdot 3 = 3 \cdot 2 = 6$, so our equation can be rewritten this way:

$75 \cdot t + 60 \cdot t = 405$

When you multiply $2 \cdot 3$, you can think of it as two threes, so our equation is really

seventy five t's plus sixty t's = 405

which is the same as

one hundred thirty five t's = 405.

(continued)

Example 6: Continued

This is really:

135 • *t* = 405.

We need a number that multiplies times 135 and becomes 405. Because 135 • 3 is 405, we see that *t* has to be equal to 3. In algebra language, we write:

t = 3

Step 4. Look Back. The number of hours for Don and Luisa to be 405 miles apart is 3. Reviewing our previous diagram, we see that this answer makes sense and checks in the original problem.

Step 3. Carry out the Plan

Carefully make certain that each step follows logically from the last. You will notice that we actually did this step in the examples for Step 2.

Step 4. Look Back

State the answer you have found and make certain that you have found what you were asked to find. Check your answer in the original problem. Now that you know the solution, can you see other ways to solve this problem? Look for other problems that you could solve with the methods you have used here. Here are four considerations for looking back.

CONSIDER THE ANSWERS THAT MAKE SENSE. What would be too much, too big, too small, too unreasonable? If you are trying to find the speed of a car, an answer of 200 mph is too big.

SOLVE THE PROBLEM ANOTHER WAY. There are many ways to solve a problem. Speculate to find other ways to check the work. Survey other people to look for new methods to increase your repertoire. Did you notice that many of the examples worked in this chapter utilized more than one strategy?

GENERALIZE. Try to make a general rule that can be applied to similar problems even though specifics change. Generalizing is what a formula does. A formula for Amy's pay from her rich uncle in Example 9 would be:

Pay on Day N in pennies = 2^{N-1}

Patterns assist in finding formulas.

REVIEW. Go back over the problem and see how the solution and attempted solutions work. Notice anything new? Do your answers make sense? If your answer is $5.00, does $5.00 work in the original problem?

REVIEW POLYA'S PROCESS

To solidify Polya's procedures in your mind, write his four steps on a paper with their strategies. Add examples to illustrate. Then practice with Polya. Try the Pushing Your Limits and Mastering Math's Mysteries. This may be a scary chapter, but your efforts here will pay off. **Go for it!**

Pushing Your Limits
CHAPTER 16

1. Write any of the Mind Shifting ideas in your journal that are possibilities for your math work. When and where might you use these? Have you used any of them before? When? How did they work? A written plan reminds you of alternatives when you are stuck.

2. List Polya's Four Steps with the problem-solving techniques in your journal. Write a short explanation or example of each. Make a photocopy to stick in your math book. As you solve math problems, determine which of these techniques you have used successfully and list any others that might also work. Practice with problems that you can already solve so that your "bag of tricks" gets bigger for future problems. (You will still need this "bag of tricks" for problem solving after you complete your math courses. You will be surprised how useful it will be for making budgets, figuring taxes, and estimating household needs.)

3. Make a web of Polya's Four Steps and the problem-solving techniques. Make up some symbols to illustrate each. Color code the steps. Fit in some examples.

4. You have started to develop your personal "bag of tricks." Push your limits with the word problems in Mastering Math's Mysteries, Chapter 16. Pick out a problem or two that appeal to you. Practice Polya's Steps, spending extra time on Steps 1 and 4. See how many of the problem-solving techniques from this section might work on one problem.

Problem Solving

Practice Polya's Four Steps and the new strategies in your "bag of tricks" by solving any of these problems. You could:

- State the goal
- Decide what is known and unknown
- Throw out irrelevant info
- Try something "off the wall"
- Guess and check
- Simplify the problem
- Act it out or use objects
- Make a picture or diagram
- Organize the data
- Identify patterns
- Work backward
- Use algebra
- Consider answers that make sense
- Or ALL OF THE ABOVE!

What a "bag of tricks"!

1. Jose loves riddles, and he has a hole in his pocket. When he pulls out all the money he has left, he reports that he has five fewer dimes than he has nickels. If the total value of his coins is now $1.60, how many nickels does he still have? *(Act this scenario out using coins. Or try to guess and check.)*

2. There are two numbers whose sum is 72. One number is twice the other. What are the numbers? *(Try to guess and check.)*

3. Jorge's rectangular garden is 15 feet longer than it is wide. Find the dimensions if the perimeter is 126 feet. *(Make a picture.)*

4. Two planes leave Orange County Airport at 1 P.M. The Delta flight heads east at 450 mph and Air Hawaii heads west at 600 mph. How long will it be before the two flights are 2,100 miles apart? *(State the goal and make a diagram.)*

5. The Santa Ana Dons drew 800 fans to their basketball game. Twenty people had free passes. Many bought student tickets for $2.00, while the remainder paid the full admission price of $3.00. Ticket sales for the game totaled $1,798.00. How many student tickets were sold? *(Simplify the problem and make a chart.)*

6. Two bikers, Katrina and Naomi, leave school at the same time. Katrina travels south at a constant rate that is 10 km/hr (kilometers per hour) faster than Naomi, who is traveling north. After five hours, they are 160 kilometers apart. Find the rate of travel (km/hr) for each cyclist. *(Draw a diagram and guess and check.)*

7. Enrique is looking for a number. He knows that three times the number minus six is equal to 45. Help him out. *(Guess and check.)*

8. Bob is anxiously waiting in his second-story office for a consultant who has an appointment at 10 A.M. It is 9:55 and Bob considers walking out to meet the man. However, he recognizes that he has no knowledge which route the man will take to his office, so he decides to sit still and wait. If there are two entrances to Bob's plant, four doors into his building, and two sets of stairs plus an elevator between the first and second floors, how many possible ways are there for the consultant to reach Bob's office? *(Throw out irrelevant data. Organize the data and make a chart.)*

17

Tackling Test Tremors

"The only thing we have

to fear is fear itself."

FRANKLIN D. ROOSEVELT

To conquer your Math Test Blues, take charge of your preparation and performance. You can know math and still not test well. Even with diligent preparation, performance anxiety can cause "static" in your brain, hindering retrieval of the knowledge you've practiced and stored.

This chapter will teach you techniques to decrease static and increase clear thinking, both before and during exams.

Remember the interrelationships between your thoughts, behaviors, emotions, and body sensations that you examined in Chapter 4. Now is a time to pay close attention to your thoughts by analyzing your Thought Distortions, neutralizing automatic negative thoughts, and choosing the best behaviors for exam preparation. This is an excellent time to call in your resources—see your instructor, join a study group, or consult a tutor.

Test preparation cannot begin the day before the test. Active preparation needs to begin a minimum of one week ahead. Better yet, you can greatly affect your test outcomes by pursuing the activities recommended in Chapters 10 through 15 of this book throughout the semester.

BEGIN WITH YOUR THOUGHTS

Here are two Interrelationship Charts (see Chapter 4) to remind you how important your thoughts and behaviors are before exams.

The Effect of Negative Thoughts

If you think to yourself, "I will fail my exam" and do not prepare, you will feel tense and fearful, and your mind will "freeze up."

If you evaluate the thought "I will fail my exam" as we did in Chapter 5, you discover it contains the Thought Distortion "Crystal Ball," because you

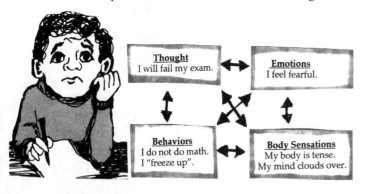

are predicting the future negatively. You can now take a deep breath and replace this thought with, "I cannot predict my test score since I am unable to see into the future." Then you begin test preparations for a positive outcome on the test. As you think more positively and actively prepare for the exam, your body relaxes and the fear calms in your mind. As you release tension and continue breathing deeply, you think more clearly and begin to enjoy the moment. See the connections on the following page.

Now let's look at other possible negative thoughts. Here is an example:

Imagine a math test is coming up. You are discouraged. You might automatically think any of the negative thoughts shown below.

Identify the Thought Distortions from Chapter 5 that are contained in these negative thoughts. Write them down now.

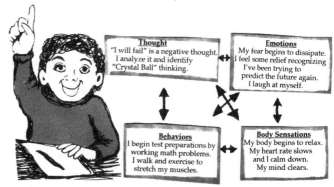

These are the Thought Distortions I identify. You might find more:

(a) Crystal Ball, Feeling = Being
(b) Mind Reading
(c) Absolute Thinking, Shoulding
(d) NEGATIVEpositive, Crystal Ball, Overgeneralization
(e) NEGATIVEpositive, Absolute Thinking, Feeling = Being
(f) Crystal Ball, Absolute Thinking, NEGATIVEpositive
(g) NEGATIVEpositive, Shoulding

Before reading on, rephrase each negative thought into a more neutral and useful statement. It must be a true statement to be effective. Then read my suggestions in the illustration on the next page.

Occasionally I feel a little discouraged and panicky when a test is coming up. That doesn't mean I won't do well. I need to prepare and can begin now by recognizing and doing problems that I know, and by clarifying what I didn't understand from today's class. The rest of this chapter will help me plan.

By next week, I will understand this new material. It is new to me now. I will practice. Understanding and confidence comes with practice.

I will choose to think and act positively as I prepare for this test. I will write down examples the teacher gives us and work them along with working review problems from the ends of the chapters in the book.

I have purposely put myself in this challenging class to grow and to enrich my life. There are many positive, concrete ways I can prepare for this test. Worrying about the test right now will not help. Worrying actually only takes energy from today. A stress-reducing walk or relaxation could help now.

Working and practicing problems will help me know more and feel better. Carving out time periods to practice, study, and ask questions will help me feel better too, and be more prepared.

Other Negative Test Thoughts

Following are other negative thoughts math students often think before exams. Put a check mark by the examples familiar to you. Identifying negative thoughts and then considering alternatives will empower you. Notice the distorted thinking as discussed in Chapter 4.

"I WILL PANIC DURING THE TEST." Unless you can foresee the future, you cannot predict your behavior or feelings. It is not uncommon to be excited. An exam is a process during which you will experience many thoughts and feelings. Actors get nervous, but they still perform. If you do panic, you can let the panic leave you. It will. No one dies from panicking during an exam.

Preparation by practicing problems, asking questions, and reviewing gives you confidence and skill that you need. Taking a dress rehearsal test and trying to panic can help you practice with out-of-control feelings. Learning relaxation techniques to use before and during exams calms you and aids clear thinking. The more you prepare ahead, the more you are in charge and feel relaxed.

"I WILL FORGET EVERYTHING." Forgetting does not mean something is gone from your mind forever. The right cue will often help you remember what you need to know. The exam will be filled with cues—words and symbols—that will trigger formulas and ideas you've practiced. Practicing strengthens the pathways in your mind.

Expecting to forget "everything" is foretelling the future and making a broad generalization. Even people with amnesia caused by illness or injury do *not* forget "everything." If you are worried about your memory, read Chapter 13.

"MATH TESTS ARE TRICKY." Math students who rely on *memorizing* the material rather than *understanding* it are usually the ones who think tests are tricky. Memorization of a few facts adds to your understanding. However, understanding basic processes and practicing to gain confidence make the "tricks" go away.

"THERE IS SO MUCH I DON'T KNOW." This will always be the case for everyone. Take a deep breath and find the level where you can begin to learn math to improve your confidence.

If you need more assistance with overwhelming negative thoughts, reread Chapters 4, 5, and 6.

PREPARATION BEFORE THE EXAM

Read on to discover how to take charge of test taking ahead of time. You are now in the driver's seat. There is much you can do to guarantee your success. As you read, check the strategies that you are willing to consider before your next exam.

1. WORK PROBLEMS. Diligently prepare and practice. The rest of this chapter cannot help during the exam if you haven't prepared your brain by learning the information and procedures for the test. This book is filled with ways to learn math with confidence.

Go over the chapters in your math book and class notes that will be covered. Look for the key ideas and skills that you have been studying. Pull out any note cards you have made with vocabulary, concepts, or problems. Look over what the instructor has told you will be on the exam. Be sure that you can do each different type of problem and be sure that you will recognize the directions so you will know when to do each procedure.

Set aside time during the week before the exam to work the chapter test or review problems at the end of each chapter. Work the examples that you copied from your instructor in class.

If you find yourself procrastinating, go to a study group or a tutor. Or make an appointment to go to the library or math study center for a specific period of time. Set short-term, immediate goals for each study session so that you are not overwhelmed and so that you feel accomplishment afterward. Take short breaks to refresh as you study.

2. LEARN THE RELAXATION RESPONSE. The "fight-or-flight" instinct automatically releases adrenaline into your body, increasing your heart rate, respiration, and metabolism to produce the energy you need to cope with emergencies. This automatic response saves people in car accidents, bear attacks, avalanches, and fires. On the down side, during an exam, the fight-or-flight response is a nuisance, preventing calm, clear thinking.

Since most math exams do not involve true emergencies such as accidents, bears, avalanches, or fires, having excessive adrenaline in your body simply makes sitting attentively and concentrating on the challenges of the test difficult.

The good news is that the antidote to the fight-or-flight syndrome was researched and documented by Dr. Herbert Benson of Harvard. He calls it the "Relaxation Response." The more the Relaxation Response is practiced, the more automatic and natural it becomes. The Relaxation Response decreases adrenaline and releases other body chemicals that promote relaxation, focus, and clear thinking. Professional athletes, musicians, and highly productive people use this technique often.

Once you learn and practice the Relaxation Response, your body relaxes easily as you simply take a deep breath at key moments during your test, including as you begin, when you get stuck, or when you need a break. Using relaxation consciously during an exam frees the thinking part of your brain

3. DO A DRESS REHEARSAL. Write your own test. Use class notes and your textbook, picking out two or more problems from each section. As you write each problem, make a separate answer key. Take your practice test in a setting similar to your exam near the same time of day. When you finish, correct your test and carefully rework the problems you missed. During the real test, you may be pleasantly surprised to find how closely you have guessed the content of the problems on the exam. See Mastering Math's Mysteries, Chapter 17, for a form that can assist you in writing your practice test.

4. ANCHOR SUCCESS TO YOUR DAY. Decide ahead of time what you will wear and what you will take to your exam. Chosen carefully, these articles will "anchor" you to yourself at your best. Be creative in providing yourself with the richest and safest possible environment so you can concentrate on the test.

The Relaxation Response

1. Sit quietly and comfortably with your eyes closed.

2. Relax your muscles, beginning with your feet and moving slowly up your body. Give each muscle group permission to "let go" as much as possible. Allow "streams of relaxation" to flow through your body.

3. Breathe slowly in through your nose, counting to five, and then breathe out through your mouth, counting to ten. If you feel dizzy, stop breathing for a moment or breathe into a paper bag.

4. As distractions come into your mind, let them go. You can take care of them later. For now, just "be" in the moment as much as possible. Simply relax as much as you can. You might imagine yourself in a safe, comfortable, and caring place just noticing what you hear, see, and feel while you are there. Any type of slow stretching, yoga, or meditation as you breathe is helpful.

5. Continue to relax and breathe for 10 to 20 minutes once or twice a day. (Do not practice these deep-breathing exercises while you drive.)

• Prepare a Test Kit to take with you to every exam. Include sharp pencils (especially your favorite ones), pens, erasers, a ruler, note cards, tissues or a handkerchief, a bottle of water, a calculator, spare batteries, etc.

• Wear clothing in which you feel good. You might wear comfortable, nonbinding clothing, or a snazzy professional-looking outfit, or your favorite "lucky shirt." Take a jacket or sweater to keep you warm in an air-conditioned classroom so oxygen-rich blood flows to your brain during the whole exam time.

• Take your "lucky rock" or "lucky pencil." Boost your confidence by pocketing a special note or card from a friend reminding you of your value to others and your life beyond this exam.

5. PREPARE YOUR BODY ALONG WITH YOUR MIND. Scientists have biological evidence that your mind is intimately intertwined with your entire body.

• Continue daily walks or bike rides or whatever aerobic exercises you use regularly. Do not neglect these stress-reducing/mind-relieving activities as you concentrate on math. If you do not have any regular exercises, choose some mild form of body movement such as stretching or walking around in your home or yard.

- Consider the foods that work best for you. Sugars and refined foods in your diet are known for reducing mental clarity. However, one or two pieces of candy (not containing chocolate) *right before* an exam has been shown to boost glucose (energy food) to the brain *temporarily*. Notice whether caffeine helps or hinders; it works for some and not for others. Protein appears to be essential for clear thinking. Carbohydrates may make you tired. Observe the foods that keep your body calm and alert and then control your food intake on test day. Pack your own food to take with you so you are not dependent on vending machines or fast-food restaurants.

- Get the optimal amount of rest. This is individual and personal, but it is important to have what works best for *you*.

6. PLAN AHEAD FOR THE TEST DAY. Make sure you know where and when your exam will be and how you will get there. Make optimal food arrangements for yourself and plan what you will wear. Pack your Test Kit. Come early to the exam to pick the best spot for yourself. Make sure that you have space around your desk so you can move if you need to.

If something happens and you find yourself rushing to your test, you may be interested to know that some studies have shown that a brisk walk before an exam has boosted test scores, probably because the increased blood circulation accelerates oxygen to the brain. Whatever happens on your test day, tell yourself that these conditions only help you concentrate more and think your best about math.

SUCCESS STRATEGIES AT THE EXAM

You can do a lot to set the tone for the day and the time of your exam. It is a special day. Excitement is natural. It is a day to speak kindly and encouragingly to yourself. It is a day to walk past any fears and to be very objective that you will do the best that you can under the circumstances. Here are some ways to make the exam a successful, even *enjoyable* experience.

1. DO A DATA DUMP. Bring a short list of formulas or facts you find difficult to remember. Look at them before the test. Visualize them going into a holding tank in your brain. Practice making them subject to recall. *If you are not allowed to use notes on the exam, be sure to put the list far away so that your honesty is not questioned.* When you receive your test, quickly write these formulas or facts on your exam paper. Now you do not have to expend any energy trying to recall them later when you need them.

2. IGNORE OTHERS BEFORE THE TEST. Do not absorb other people's anxiety like a sponge. Some people deal with their anxiety by verbally expressing their worst fears and concerns in an excited or hysterical way. If these conversations bother you, protect yourself by building an invisible shield around yourself. Merely observe their anxious state.

It is perfectly O.K. to avoid talking with anyone and to sit by yourself before the exam so that you can breathe deeply and focus. This is a time for you to care for yourself in whatever way you need to. Choose what you do and whom you speak with before each exam. You will feel more in charge.

3. BREATHE AND RELAX. When you feel stuck or tense, take a deep breath in through your nose and out through your mouth. Let "everything" go as you expel the air. (The more you have practiced the Relaxation Response before the test, the more you will be able to relax during the test.) Massage your temples, scalp, and the back of your neck to increase blood flow with oxygen to your brain to help you think more clearly.

4. TAKE TIME OUT. Take short breaks during the exam to close your eyes, breathe deeply and stretch your neck and arms. Tell yourself that you have all the time that you will need. A few isometric exercises can release tension too. With your feet flat on the floor, take hold of the sides of your chair and pull up. Close and relax your eyes by covering them for a short time with your cupped palms. Draw a small empty box on your paper. Imagine it is a TV screen and picture a wonderful, relaxing scene.

5. USE YOUR SUBCONSCIOUS MIND. If a problem makes no sense, read it and go on. Ideas will come to you as the problem sinks into your subconscious mind and you continue with the test. You may wish to read over any essay questions or word problems early in the exam to let your subconscious mind begin to understand and process them, collecting ideas as you work the other problems.

6. TRUST. Let each question reach into your mind for the answer. Remind yourself that you know everything you need to know for now.

7. STRATEGIZE. Skim the test. You do not have to take the exam in the order it appears. Do the problems and questions that you like first. Make tiny pencil marks by those questions to which you want to return. Don't watch the clock more often than necessary. The instructor will stop you when the exam is over.

If some of the questions are multiple choice, eliminate the obvious wrong answers first and then do the work until you can choose the exact answer. Be sure to simplify your answer.

8. USE TIME WISELY. Don't work on one problem for a long time. Often a question further into the exam will act as a "key" to unlock a previous problem. Tell yourself that you have all of the time you need. Let go of the rest of your life during the exam. You can deal with all that later.

9. IGNORE OTHER PEOPLE DURING THE TEST. You do not need to compete with anyone except yourself. No matter how many pages anyone else turns or how much noise they make or how soon they leave the exam, you will not know how well they performed. Don't compare. You need this energy to do your work. Often the first person to leave an exam gets a very low score. Ignore him. Often the last person to leave gets a very high score. Take your time. Let other people's behavior go for now.

Earplugs help shut out excess noise. If you use them, notify the instructor so that you will not miss any announcements. Sit where you can't see others. If possible, keep your desk clear of touching other desks so you are not disturbed. If you notice a distraction, allow yourself to notice it and then let it go. Allow the instructor to take care of problems in the room. You need only take charge of yourself and your performance right now.

10. ASK QUESTIONS. Ask the instructor questions as needed. Raise your hand and keep working. Let the instructor come to you. The worst thing that can happen is that the instructor will say, "I can't answer that question." Often just giving yourself permission to ask a question allows you to think differently and to figure it out on your own.

REFOCUS AFTER THE EXAM

1. AFTER THE EXAM, LET THE RESULTS GO. You have used a lot of energy and may be low and off balance. You may wish to pass up discussing the exam with others so you can take care of yourself. Going to the bathroom, drinking some water, and eating something can help you feel normal again. You may have put much of your life aside to prepare for this exam. Refresh yourself and get your life back. You can deal with the test results later when your priorities are in order again.

2. WHEN YOU GET YOUR TEST RESULTS, STAY CALM. Do not compare results with anyone else. These results are between you and the instructor. A good grade does not mean you are a wonderful, great person any more than a poor grade means that you are a horrible person. The test was just a one-time evaluation of how you were able to do certain problems with the understanding you had at the time. If need be, you can learn the material and repeat the exam or the course. Test results are not necessarily fun, but they are not the end of the world *and* they mean nothing about your worth as a person.

3. KEEP THE EXAM AND THE ANSWERS IF YOU CAN. If you feel emotionally flooded by your test results, give yourself some recovery time. Later, when you have been able to bring some perspective to the situation, review your exam and learn the material that you missed. This is a great opportunity for you to get feedback about your learning.

Notice the kinds of mistakes that you made.

- Which errors were caused by your not knowing the material from lack of studying?
- Which errors were caused by your misunderstanding a concept or procedure?
- Which errors were caused by your not remembering something that you learned?
- Which errors were caused by your misreading or not reading the directions?

Knowing the types of errors that you make gives you direction for preparation next time. Since new math material depends upon what you've covered previously, spending quality time with old exams will pay you big dividends. Keep your old exams for practice and review. They will be especially useful two weeks before your final exam.

4. ANSWER THE FOLLOWING QUESTIONNAIRE. Completing this evaluation will help you objectively evaluate your exam experience and prepare for the next. Be honest—your answers will give you a reality check about what you can do differently to obtain higher scores. Remember that it is not how smart you are, it is whether or not you are persistent and do what needs to be done to learn the math.

Exam Evaluation

- ❏ Did you do the assigned homework for the test material?
- ❏ Did you attend every class session before the test?
- ❏ Were you on time to class and prepared with your paper, pencil, and textbook when class began?
- ❏ Did you take thorough class notes, copying down what the instructor wrote and said, including all the examples?
- ❏ Did you complete your homework as soon after class as possible?
- ❏ Did you write a practice dress rehearsal test, take it, and correct it before the exam?
- ❏ Did you ask questions on homework problems or concepts that you did not understand?

(continued)

Exam Evaluation Continued

❏ Did you have a regular time and place to do your math studying?

❏ Did you use the tutoring services on campus?

❏ Did you make a web in preparation for this exam?

❏ Did you actually study for the exam by working problems from the book and your notes?

❏ Did you practice the test-taking strategies suggested in this chapter?

❏ Did you consult your instructor, tutor, or fellow math students when you needed outside input or assistance?

❏ Did you take care of your body by eating nutritiously and getting sufficient rest during the week before and the day of the test?

❏ Did you practice the Relaxation Response in preparation for the exam?

❏ Did you consciously take deep breaths and relax during the exam?

❏ Did you choose your classroom seat to avoid distractions?

❏ Are you satisfied with this test score?

These questions point to possible changes you can make before your next exam. What actions will you take this week to ensure a higher score next time?

Pushing Your Limits

17
CHAPTER

1. Write down three specific things that you will do differently *before* your next exam. How and when will you do them? After the exam, reread what you have written and evaluate your plan.

2. Write down three specific things that you will do *at* your next exam. Plan carefully how to execute these strategies. How will you remember what to do? Evaluate your success after the test and refine your plan.

3. Which of the negative thoughts seem familiar to you? Write them down and then rewrite them in a neutral, alternative way. Write some other related alternative thoughts.

4. Answer the exam evaluation after the exam. What three things can you do to raise your test results?

5. Web the contents of this chapter. Highlight the techniques you are willing to explore.

6. Mastering Math's Mysteries, Chapter 17, will guide you to write a practice test with an answer key. Choose problems from your notes and textbook that have answers you can use to correct yourself. Take your practice test at a time and in an environment similar to your testing situation. Check the test and rework the problems. How did it feel? Were you prepared for your exam?

Dress Rehearsal Test

1. Choose sections of your math textbook for a practice exam for your next class test. Your teacher or your class syllabus can help you choose the appropriate sections.

2. From each section of your math textbook, copy problems 7 and 37 (or 17 or 27) along with the directions onto the following form. These two problems will give you a sampling of practice problems. (You may have more or fewer than 20 problems, depending on the number of sections your test will cover.)

3. From the back of your textbook, copy the answers to the odd problems you chose onto the answer key on page 210.

4. Set aside time to take this practice test two days before your exam so that you have time to correct it and get the help that you need for success.

   (*Note:* If you are not in a math class now, write a dress rehearsal exam over the Mastering Math's Mysteries worksheets that you have worked in this book. Be sure to copy the directions for each problem and the solutions.)

1.

2.

3.

4.

5.

6.

7.

8.

9.

10.

11.

12.

13.

14.

15.

16.

17.

18.

19.

20.

Dress Rehearsal Test Answer Key

1.	2.
3.	4.
5.	6.
7.	8.
9.	10.
11.	12.
13.	14.
15.	16.
17.	18.
19.	20.

Take Charge of Success

"The only way to discover the limits of the possible is to go beyond them into the impossible."

ARTHUR C. CLARKE

The key to progress with math is consciously taking charge of your thoughts and actions. Then you control your work with math instead of letting it control you, and you take charge of your success. Tape these recommendations in your math book.

1. ASK QUESTIONS. It is the smart thing to do. Be active and assertive. Learning is not a spectator sport. You can't learn well from the sidelines. Get involved. Work problems and keep asking questions until they become clear. In class, ask questions on confusing procedures.

2. HANG IN. When you get stuck working problems, hang in for a while and then take a break. Go back later, and begin at the beginning with a clean sheet of paper and a different point of view. Just because you don't understand at first does not mean understanding won't come. Math learning requires time to settle into your brain. Being able to live with uncertainty for a while is a good math skill to have.

3. KEEP A LIST. Write down your resources (instructors, other students, books, tutors, people to answer questions, people who understand) so that you can consult them when you get discouraged. You are not alone. Find helpful people with whom you are comfortable. Form a network of others working toward the same goals as you. Use your network!

4. FIND YOURSELF. Discover your own unique ways of learning. Experiment with new ones. If a method doesn't work, find others. Ask different people how they learn math or do a problem. They will often feel honored and pleased that you asked them, and you might get a breakthrough idea.

5. BE POSITIVE. Listen to what you say to yourself inside your head. It is difficult to work well if you are saying, "I'll never get this" or "I can't do math." Change those negative messages to neutral ones like "I haven't gotten this *yet*" or "I can't do this particular problem *yet.*"

6. TAKE A BREAK. Use any relaxation techniques that you know. Take a deep breath or stretch occasionally.

7. LEARN FROM MISTAKES. Remember that errors are part of the learning process. Pay attention to them and figure out where they happened and how to fix them.

8. KEEP IT REAL. Be realistic with your expectations of yourself—your math level, your life commitments, your time constraints. Don't beat on yourself for being a human being.

9. USE TECHNOLOGY. Learn to use a calculator and use it *appropriately* for calculations with large numbers and decimals. Each brand of calculator is different, so keep your manual for reference. Take spare batteries to exams.

10. START EASY. Practice the easier math problems to warm up each time you begin your math study. This builds confidence and strengthens those math pathways in your brain.

11. USE PAPER. Keep scratch paper available and expect to use it for your math work. You need empty space on paper to think and do calculations.

12. PROMOTE EMOTIONAL WELL-BEING. Patience, self-care, and humor will make your math work so much easier. Your brain will work better too.

13. BE HEALTHY. You are making new connections in your brain as you practice math, so sufficient sleep and exercise plus healthy foods are important. Having fresh drinking water available and breathing fresh air also help you think better.

14. REWARD YOURSELF. Acknowledge your progress—every little bit! Pat yourself on the back for each and every problem you work. Notice what you know now that you didn't know two weeks ago. Document your growth in your journal.

I wish you well.

APPENDIX

Mean Math Blues Boogie

(See the following page for accompanying music.)

Got the Mean Math Blues since I shined on my home-work.
Got the Mean Math Blues ever since I tried this perk.
I thought math too easy.
I thought I could do it.
I thought that I followed—
There was nothin' to it.
Got the Mean Math Blues ever since I skipped my work.

Got the Mean Math Blues ever since I didn't ask.
Got the Mean Math Blues ever since I slipped this task.
Had a little question—
Didn't stop to ask it.
Now I have confusion—
Better not to mask it.
Got the Mean Math Blues ever since I didn't ask.

Got the Mean Math Blues ever since I missed math class.
Got the Mean Math Blues ever since I took a pass.
Got up on the wrong side.
Didn't want to face math.
Class went on without me.
Guess I took the wrong path.
Got the Mean Math Blues ever since I missed my class.

Mean Math Blues Boogie

Bibliography

Introduction

Bailey, Elinor Peace. "Let's Face It." Piecemakers, Costa Mesa, CA. 18 April 1999.

Bailey, Elinor Peace. *Mother Plays with Dolls . . . and finds an important key to unlocking creativity*. McLean, VA: EPM Publications, 1990.

Cooney, Miriam P. csc, ed. *Celebrating Women in Mathematics and Science*. Reston, VA: National Council of Teachers of Mathematics, 1996.

Folsing, Albrecht. *Albert Einstein*. New York: Viking Penguin, 1997.

Kass-Simon, G., and Patricia Farnes, ed. *Women of Science: Righting the Record*. Bloomington, IN: Indiana University Press, 1990.

Maloof, Sam. *Sam Maloof Woodworker*. New York: Kodansha America, 1983.

McGrayne, Sharon Bertsch. *Nobel Prize Women in Science: Their Lives, Struggles and Momentous Discoveries*. New York: Birch Lane Press, 1993.

Ooten, Cheryl. Address. Santa Ana College Commencement. Santa Ana Stadium, Santa Ana, CA. 4 June 1999.

Polster, Miriam. *Eve's Daughters: The Forbidden Heroism of Women*. San Francisco: Jossey-Bass, 1992.

Chapter 1

Beck, Aaron, M.S., and Gary Emery, Ph.D. *Anxiety Disorders and Phobias: A Cognitive Perspective*. New York: Basic Books, 1985.

Goldberg, Natalie. *Writing Down the Bones*. Boston: Shambhala, 1986.

Greenberger, Dennis, Ph.D., and Christine A. Padesky, Ph.D. *Mind Over Mood: Change How You Feel by Changing the Way You Think*. New York: Guilford Press, 1995.

LeDoux, Joseph. *The Emotional Brain*. New York: Touchstone, 1996.

Tobias, Sheila. *Overcoming Math Anxiety*. New York: W. W. Norton & Co., 1993.

Chapter 2

Hackworth, Robert D. *Math Anxiety Reduction*, 2nd ed. Clearwater, FL: H&H Publishing Co., 1992.

Chapter 3

Adair, Margo. *Working Inside Out: Tools for Change*. Berkeley: Wingbow Press, 1984.

Bandler, Richard. *Using Your Brain for a Change*. Moab, UT: Real People Press, 1985.

Bandler, Richard, and John Grinder. *Frogs into Princes*. Moab, UT: Real People Press, 1979.

Gardner, Howard. *Multiple Intelligences: The Theory in Practice*. New York: Basic Books, 1993.

Gross, Ronald. *Peak Learning*. New York: Tarcher/Putnam, 1991.

Kogelman, Stanley, and Barbara R. Heller. *The Only Math Book You'll Ever Need*. New York: HarperCollins, 1994.

Kogelman, Stanley, and Joseph Warren. *Mind over Math*. New York: McGraw-Hill, 1978.

Tobias, Sheila. *Overcoming Math Anxiety*. New York: W. W. Norton & Co., 1993.

Chapter 4

Beck, Aaron, M.D., and Gary Emery, Ph.D. *Anxiety Disorders and Phobias: A Cognitive Perspective.* New York: Basic Books, 1985.

Burns, David D., M.D. *Feeling Good: The New Mood Therapy.* New York: Avon, 1999.

Greenberger, Dennis, Ph.D., and Christine A. Padesky, Ph.D. *Mind over Mood.* New York: Guilford Press, 1995.

Pert, Candace, Ph.D. *Molecules of Emotion.* New York: Scribner, 1997.

Sapolsky, Robert, Ph.D. "Stress, Disease and Memory," Brain Expo 2000. Paradise Point Hotel, San Diego. 18 January 2000.

Siegel, Daniel, M.D. "The Developing Mind," Brain Expo 2000. Paradise Point Hotel, San Diego. 18 January 2000.

Chapter 5

Beck, Aaron, M.D., and Gary Emery, Ph.D. *Anxiety Disorders and Phobias: A Cognitive Perspective.* New York: Basic Books, 1985.

Burns, David D., M.D. *Feeling Good: The New Mood Therapy.* New York: Avon, 1999.

Greenberger, Dennis, Ph.D., and Christine A. Padesky, Ph.D. *Mind over Mood.* New York: Guilford Press, 1995.

Sapolsky, Robert, Ph.D. "Stress, Disease and Memory," Brain Expo 2000. Paradise Point Hotel, San Diego. 18 January 2000.

Chapter 6

Adair, Margo. *Working Inside Out: Tools for Change.* Berkeley: Wingbow Press, 1984.

Beck, Aaron, M.D., and Gary Emery, Ph.D. *Anxiety Disorders and Phobias: A Cognitive Perspective.* New York: Basic Books, 1985.

Burns, David D., M.D. *Feeling Good: The New Mood Therapy.* New York: Avon, 1999.

Cameron, Julia. *The Artist's Way.* New York: Tarcher/Putnam, 1992.

Greenberger, Dennis, Ph.D., and Christine A. Padesky, Ph.D. *Mind over Mood.* New York: Guilford Press, 1995.

Gross, Ronald. *Peak Learning.* New York: Tarcher/Putnam, 1991.

Chapter 7

Clawson, Calvin C. *Mathematical Mysteries: The Beauty and Magic of Numbers.* New York: Plenum Press, 1996.

Enzensberger, Hans Magnus. *The Number Devil: A Mathematical Adventure.* New York: Metropolitan Books, 1997.

Gardner, Howard. *Multiple Intelligences: The Theory in Practice.* New York: Basic Books, 1993.

Gullberg, Jan. *Mathematics: From the Birth of Numbers.* New York: W. W. Norton & Co., 1997.

Nelson, David, George Gheverghese Joseph, and Julian Williams. *Multicultural Mathematics.* New York: Oxford University Press, 1993.

Shorris, Earl. *Latinos: A Biography of the People.* New York: Avon Books, 1992.

Sunbeck, Deborah. *Infinity Walk: Preparing Your Mind to Learn!* Torrance, CA: Jalmar, 1996.

Chapter 8

Bandler, Richard. *Using Your Brain for a Change.* Moab, UT: Real People Press, 1985.

DePorter, Bobbi, and Mike Hernacki. *Quantum Learning: Unleashing the Genius in You.* New York: Dell, 1992.

Sunbeck, Deborah. *Infinity Walk: Preparing Your Mind to Learn!* Torrance, CA: Jalmar, 1996.

Chapter 9

Albers, Donald J., Gerald L. Alexanderson, and Constance Reid, ed. *More Mathematical People: Contemporary Conversations.* San Diego: Academic Press, 1990.

Clawson, Calvin C. *Mathematical Mysteries: The Beauty and Magic of Numbers.* New York: Plenum Press, 1996.

Enzensberger, Hans Magnus. *The Number Devil: A Mathematical Adventure.* New York: Metropolitan Books, 1997.

Kass-Simon, G., and Patricia Farnes, ed. *Women of Science: Righting the Record*. Bloomington, IN: Indiana University Press, 1990.

McGrayne, Sharon Bertsch. *Nobel Prize Women in Science: Their Lives, Struggles, and Momentous Discoveries*. New York: Birch Lane Press, 1993.

Morrison, Philip, Phylis Morrison, and The Office of Charles and Ray Eames. *Powers of Ten*. New York: Scientific American Library, 1994.

Pappas, Theoni. *The Magic of Mathematics*. San Carlos, CA: Wide World Publishing/Tetra, 1996.

Polster, Miriam. *Eve's Daughters: The Forbidden Heroism of Women*. San Francisco: Jossey-Bass, 1992.

Reid, Constance. *Julia: A Life in Mathematics*. Washington, DC: Mathematical Association of America, 1996.

Stewart, Ian. *Another Fine Math You've Got Me Into . . .* New York: W. H. Freeman, 1992.

Tahan, Malba. *The Man Who Counted*. New York: W. W. Norton & Co., 1993.

Chapter 10

Cleary, Beverly. *Ramona the Pest*. New York: William Morrow, 1968.

Csikszentmihalyi, Mihaly. *Finding Flow*. New York: Basic Books, 1997.

DePorter, Bobbi, and Mike Hernacki. *Quantum Learning: Unleashing the Genius in You*. New York: Dell, 1992.

Engelbreit, Mary. Card. Bloomington, IN: SUNRISE, Inc., 1996.

Hart, Leslie. *Human Brain & Human Learning*. Kent, WA: Books for Educators, 1998.

Chapter 11

Andreas, Brian. *Mostly True*. Decorah, IA: StoryPeople, 1993.

Hart, Leslie A. *Human Brain & Human Learning*. Kent, WA: Books for Educators, 1998.

Schrof, Joannie M., and Stacey Schultz. "Social Anxiety." *U.S. News & World Report*, 21 June 1999.

Sousa, David, Ed.D. "Seven Critical Discoveries about How the Brain Learns," Brain Expo 2000. Paradise Point Hotel, San Diego. 18 January 2000.

Wolfe, Pat. "Lessons Learned from Brain Research," Brain Expo 2000. Paradise Point Hotel, San Diego. 18 January 2000.

Chapter 12

McMeekin, Gail. *The 12 Secrets of Highly Creative Women*. New York: MJF Books, 2000.

Chapter 13

Cahill, Larry, Ph.D. "Emotions and Memory." Brain Expo 2000, Paradise Point Hotel, San Diego. 19 January 2000.

Grafton, Sue. *"O" is for Outlaw*. Markham, Ontario, Canada: Henry Holt, 1999.

Markowitz, Karen, M.A., and Eric Jensen, M.A. *The Great Memory Book*. San Diego: The Brain Store, 1999.

Rossi, Ernest Lawrence, Ph.D., with David Nimmons. *The 20 Minute Break*. Los Angeles: Jeremy P. Tarcher, 1991.

Sapolsky, Robert, Ph.D. "Stress, Disease, and Memory." Brain Expo 2000. Paradise Point Hotel, San Diego. 18 January 2000.

Siegel, Daniel, M.D. "The Developing Mind." Brain Expo 2000. Paradise Point Hotel, San Diego. 18 January 2000.

Zola, Dr. Stuart. "Memory Seminar." Institute for Cortext Research and Development Seminars. Marriott Hotel, Anaheim, California. 20 September 1999.

Chapter 14

Bromley, Karen, Linda Irwin-DeVitis, and Marcia Modlo. *Graphic Organizers: Visual Strategies for Active Learning*. New York: Scholastic, 1995.

Buzan, Tony, with Barry Buzan. *The Mind Map Book: How to Use Radiant Thinking to Maximize Your Brain's Untapped Potential.* New York: Dutton, 1994.

Wycoff, Joyce. *Mindmapping*®: Your Personal Guide to Exploring Creativity and Problem-Solving. New York: Berkley, 1991.

Chapter 15

Student/Tutor Interviews, Santa Ana College, June 2001.

Chapter 16

Afflack, Ruth. *Beyond Equals: To Encourage the Participation of Women in Mathematics.* Oakland, CA: Math/Science Network, 1982.

Enzensberger, Hans Magnus. *The Number Devil: A Mathematical Adventure.* New York: Metropolitan Books, 1997.

Kogelman, Stanley, and Barbara R. Heller. *The Only Math Book You'll Ever Need.* New York: HarperCollins, 1994.

Moretti, Gloria, et al. *The Problem Solver 6: Activities for Learning Problem-Solving Strategies.* Mountain View, CA: Creative Publications, 1987.

Polya, George. *How to Solve It.* 2nd Ed. Princeton, NJ: Princeton University Press, 1988.

Rossi, Ernest Lawrence, Ph.D., with David Nimmons. *The 20 Minute Break.* New York: Jeremy P. Tarcher, Inc., 1991.

Stone, Irving. *The Origin.* New York: Doubleday, 1980.

Chapter 17

Benson, Herbert, M.D. *The Relaxation Response.* New York: Avon, 1975.

Burns, David D., M.D. *Feeling Good: The New Mood Therapy.* New York: Avon, 1999.

Appendix

Afflack, Ruth. *Beyond Equals: To Encourage the Participation of Women in Mathematics.* Oakland, CA: Math/Science Network, 1982.

More Mastering Math's Mysteries

I have put more challenging math exercises in this section for those math students who are beyond the basic levels. You may or may not be able to work these mysteries alone. Do not be dismayed. Teaming up with another math student, a tutor, or a teacher is a great way to learn and build professional skills.

THE PASCAL TRIANGLE

The following arrangement of numbers probably was discovered when a mathematician was multiplying an algebra expression like "$x + 1$" times itself over and over again. At this point in your math career, you do not need to do the multiplication or to understand how the numbers came about from it, *but* you can certainly look at the triangle, called the Pascal Triangle, that resulted and be fascinated by it. It includes many, many patterns.

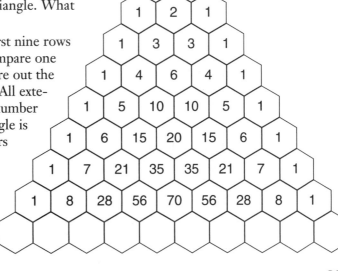

1. Examine the numbers on both the left side and the right side of the triangle. What pattern do you notice?

2. I have included only the first nine rows of the Pascal Triangle. Compare one row with the next and figure out the numbers in row 10. (*Hint:* All exterior numbers are 1. Each number on the interior of the triangle is the sum of the two numbers diagonally above it.)

3. Use a straightedge to draw a vertical line down the center of the triangle. What symmetry do you see

Pascal Triangle (partially filled):

```
            1
          1   1
        6
      5
          6
        35
    8  28  56  70  56  28   8
  1  9  36  84  126  126  84  36  9  1
```

on the right and left sides? Do you see how the numbers on one side mirror those on the other?

4. Use the patterns you found in #2 to fill in the blanks on the triangle to the left. Fill in the exterior numbers first. Then, starting at the top, work down to fill in the interior numbers. Check your work with the previous page.

5. Photocopy two copies of the Pascal Triangle.

   • On one copy, color the counting numbers: 1, 2, 3, 4, . . . and then, with a different shade, color the Triangular Numbers from Mastering Math's Mysteries, Chapter 5.

   • On another copy of the triangle, shade in all of the even numbers. If you added more rows to your triangle, this would look even more interesting!

6. Add the numbers in each row. Row 1 adds up to 1. Row 2 adds up to 2, etc.

   Record the totals here: 1, 2, 4, ____, ____, ____, ____, ____, . . .

   What is the pattern in this sequence of numbers?

7. *Optional:* If you are very brave, find a higher-level math student or a tutor or teacher who is willing to give you the answers to the following algebra multiplication problems. Ask for the answers "in descending order."

   (a) $(x + 1)^0$

   (b) $(x + 1)^1$

   (c) $(x + 1)^2$

   (d) $(x + 1)^3$

   (e) $(x + 1)^4$

   • When you see the answers, ignore the letters with their exponents. (In x^5, 5 is called the exponent, or power.)

   • Only look at the numbers to the left of the letters. If there is no number left of the letter, as with x^5, the number is understood to be 1.

   • Compare the answer from (a) above to Row 1 of the Pascal Triangle.

- Compare the answer from (b) above to Row 2 of the Triangle.
- Compare the answer from (c) above to Row 3 of the Triangle.
- And so on. What do you notice? Do you see where the triangle's numbers came from?
- Look for patterns in the letters too. Do you think that you could write down the answer to $(x + 1)^5$ without knowing algebra and multiplying? You could do it if you discover the patterns and use the Pascal Triangle!

FIBONACCI AND PASCAL

There are many more patterns that appear in the Pascal Triangle that we haven't discussed. However, I do want you to see that the Fibonacci sequence is there too. Seeing it is a bit tricky. Below, I have turned the diagram of the Pascal Triangle and drawn a series of horizontal lines through it.

- Lay a piece of paper over the triangle so that the top edge of the paper coincides with the top horizontal line that I have drawn. How did I get the 1?
- Move your paper down one line. How did I get 1 again?
- Move your paper down again so the top edge matches the third horizontal line. Notice that 1 + 1 is 2.
- Continue moving the paper down. Add the numbers in the hexagons (the six-sided figures surrounding each of the numbers) that appear directly above the top edge of your paper.

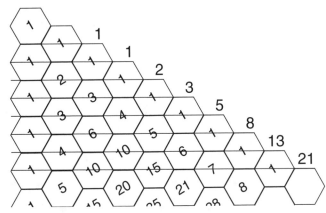

I really hope that you are able to find the Fibonacci sequence in the Pascal Triangle. I find it quite intriguing. Fibonacci is everywhere!

PICK'S THEOREM

In 1899, over a century ago, B. Pick discovered a unique method (Afflack, 1982) to find the area of rectangles, squares, parallelograms, trapezoids, triangles, and more. You can replicate his method using Geopaper (the dot paper at the end of this exercise), a straightedge, and a sharp pencil.

PICK'S METHOD

1. Draw any figure with straight sides and no holes on Geopaper. Use a straightedge and a sharp pencil. You can use a Geoboard and rubber bands if they are available.
2. Count the number of Boundary Points—those Geopaper points that actually lie on the figure. Call that number *B*.
3. Count the number of Interior Points—those Geopaper points that lie inside the figure. Call that number *I*.
4. To calculate the area, divide *B* by 2, then add I and subtract 1.

Pick's Theorem (in shorthand): $A = \frac{B}{2} + I - 1$

Let's try an example. In this figure, *B* = 11 and *I* = 5 (marked by *x*). Dividing 11 by 2, we get $5\frac{1}{2}$. Adding $5\frac{1}{2}$ to 5 and subtracting 1, the area is $9\frac{1}{2}$ square units.

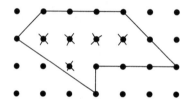

1. Find the area of the following figures (a through g) using Pick's Theorem. Record your results in the t-table on the next page.

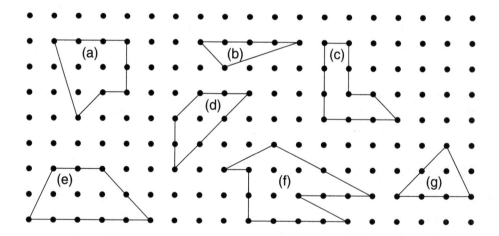

	BOUNDARY POINTS	INTERIOR POINTS	AREA
(a)			
(b)			
(c)			
(d)			
(e)			
(f)			
(g)			

2. Photocopy the Geopaper reproduced on page 226. Sketch each figure described below on Geopaper and use Pick's Theorem to find the area (in square units).

FIGURE	BASE(S)	HEIGHT	AREA
(a) Triangle	10 ft	3 ft	
(b) Parallelogram	10 ft	3 ft	
(c) Trapezoid	6 m & 4 m	5 m	
(d) Rectangle	5 cm	3 cm	
(e) Square	7 inches	7 inches	
(f) Trapezoid	5 yd & 3 yd	4 yd	
(g) Parallelogram	12 m	4 m	
(h) Triangle	12 m	4 m	
(i) Rectangle	11 ft	5 ft	
(j) Trapezoid	3 cm & 9 cm	2 cm	
(k) Parallelogram	6 in	5 in	

Answer Key

SOLUTIONS TO MASTERING MATH'S MYSTERIES

Chapter 3, Counting

1. 0, 1, 2, 3, 4, 5, 6, 7, 8, 9
2. 16, 18, 20
3. 6, 9, 12, 15, 21, 24, 27
4. 8, 12, 16, 20, 24, 32, 36
5. 10, 15, 20, 25, 30, 40, 45, 50
6. 12, 18, 24, 36, 42, 48, 60

Chapter 4, Patterns and Square Numbers

1. 5 • 5, 6 • 6, 8 • 8, 9 • 9, 11 • 11, 12 • 12, 16, 25, 100
2. 4, 16, Square that is 5 units on each side, Square that is 6 units on each side
3. 1, 4, 9, 16, 25, 36, 49, 64, 81
4. 144, 169, 196, 225, 256, 289, 324, 361, 15 • 15, 16 • 16, 17 • 17, 18 • 18, 19 • 19
5. 5^2, 6^2, 8^2, 9^2, 9, 100

Chapter 5, Triangular Numbers

1. 1 + 2 + 3 + 4 = 10, 1 + 2 + 3 + 4 + 5 = 15
2. 6, A triangle of 21 dots with 6 dots in the bottom row
3. 21, 28, 36, 45, add 5, add 11
4. Count the squares in each "staircase" to get a Triangular Number. Fifteen is a stairway of blocks with 5 in the right column. Twenty-one is a stairway of blocks with 6 in the right column.

Chapter 6, Fibonacci Numbers

1. 21, 34, 55, 89
2. 1, 2, 1.5, 1.666..., 1.6, 1.625, 1.615, 1.619

Chapter 7, Adding & Subtracting Fractions

1. $\frac{1}{10}$
2. $\frac{5}{37}$
3. $\frac{5}{8}$
4. $\frac{11}{12}$
5. $\frac{10}{x}$
6. $\frac{5}{12}$

Chapter 8, Working with Halves

1. $3\frac{1}{2}$
2. $4\frac{1}{2}$
3. $1\frac{1}{2}$
4. $1\frac{1}{2}$
5. $7\frac{1}{2}$
6. $2\frac{1}{2}$
7. $3\frac{1}{8}$
8. $3\frac{3}{5}$

Chapter 9, More on Fraction Adding & Subtracting

1. 2, 4, 6, 8, 10
2. $\frac{7}{12}$
3. $\frac{10}{12}$ or $\frac{5}{6}$
4. $\frac{6}{12}$ or $\frac{1}{2}$
5. $\frac{5}{12}$
6. $\frac{7}{12}$
7. $\frac{5}{12}$
8. $\frac{7}{12}$
9. $\frac{1}{12}$
10. $\frac{1}{6} + \frac{1}{3} + \frac{1}{4}$ is $\frac{9}{12}$ or $\frac{3}{4}$

Chapter 10, Multiplying Fractions

1. $\frac{3}{36}$ or $\frac{1}{12}$
2. $\frac{1}{6}$
3. $\frac{3}{12}$ or $\frac{1}{4}$
4. $\frac{5}{30}$ or $\frac{1}{6}$
5. $\frac{3}{36}$ or $\frac{1}{12}$
6. $\frac{6}{12}$ or $\frac{1}{2}$
7. $\frac{2}{24}$ or $\frac{1}{12}$
8. $\frac{5}{12}$

Chapter 11, Dividing Fractions

1. $\frac{2}{3} \bullet \frac{3}{2} = \frac{6}{6}$ *or* 1
2. $\frac{3}{4} \bullet \frac{4}{1} = \frac{12}{4}$ *or* 3
3. $\frac{1}{4} \bullet \frac{12}{1} = \frac{12}{4}$ *or* 3
4. $\frac{5}{6} \bullet \frac{12}{1} = \frac{60}{6}$ *or* 10
5. $\frac{1}{2} \bullet \frac{6}{1} = \frac{6}{2}$ *or* 3
6. $\frac{4}{5} \bullet \frac{10}{2} = \frac{40}{10}$ *or* 4

Chapter 12, Sum, Difference, Product, Quotient

1. 11
2. 22
3. 4
4. 11
5. 36
6. 24
7. 6
8. 3
9. 21
10. 17
11. $\frac{3}{7}$
12. 4

13. 4   14. 72   15. 7   16. 21

17. $\frac{4}{15} + \frac{9}{15} = \frac{13}{15}$ 18. $\frac{2}{3} \div \frac{1}{3} = \frac{2}{3} \cdot \frac{3}{1} = \frac{6}{3}$ *or* 2

19. $\frac{3}{8} \cdot \frac{1}{3} = \frac{3}{24}$ *or* $\frac{1}{8}$ 20. $\frac{7}{12} - \frac{2}{12} = \frac{5}{12}$

Chapter 13, Memory Devices for Fractions

1. $\frac{6}{7}$ 2. $\frac{5}{x}$ 3. $\frac{x+3}{9}$ 4. $\frac{12}{y}$

5. $\frac{3-x}{7}$ 6. $\frac{5+x}{14z}$ 7. $\frac{8}{49}$ 8. $\frac{3x}{28}$

9. $\frac{8}{15}$ 10. $\frac{7}{2y}$ 11. $\frac{18}{25}$ 12. $\frac{9}{35}$

13. $\frac{7}{20}$ 14. $\frac{2x}{15}$ 15. $\frac{4}{9}$ 16. $\frac{35}{7}$ or 5

17. $\frac{15}{12}$ or $\frac{5}{4}$ or $1\frac{1}{4}$ 18. $\frac{3}{3}$ or 1

Chapter 14, Webbing

Answers will vary.

Chapter 15, Skills to Bridge the Gaps

3.

(a)

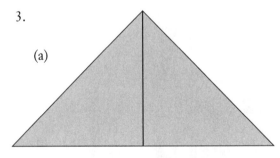

(b)

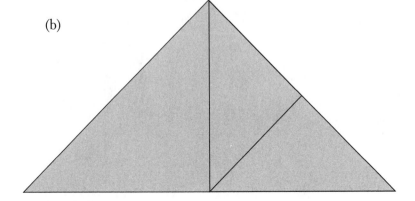

4.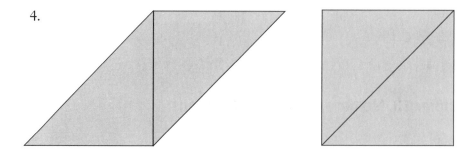

5. To make the medium triangle, match the short sides with right angles together. To make the parallelogram, match short sides but not right angles. To make the square, match the long sides.

6.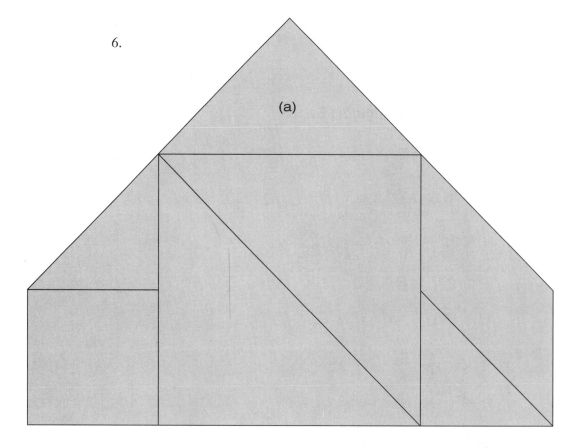

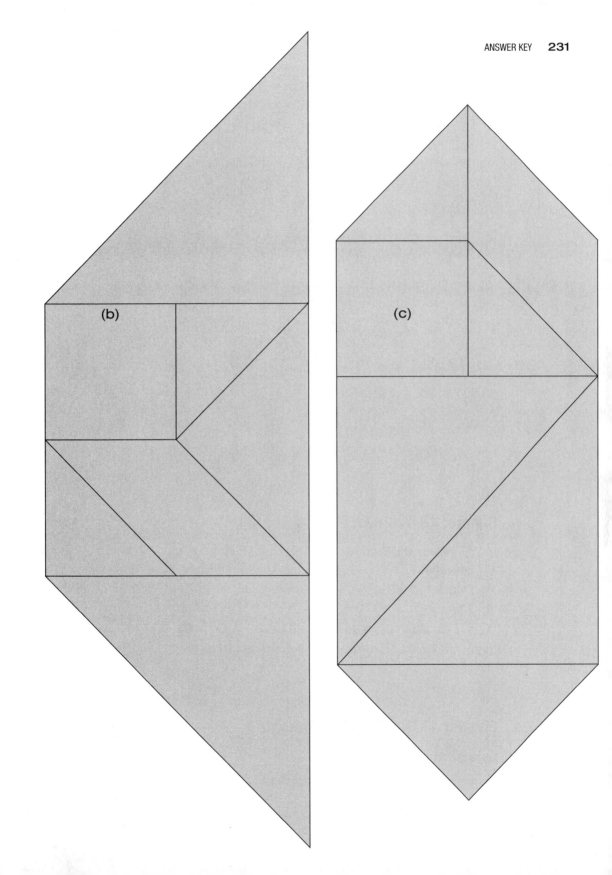

7.

Chapter 16, Problem Solving

1. 14 nickels and 9 dimes
2. 24 and 48
3. 24 feet by 39 feet
4. 2 hours
5. 542 students tickets and 238 full tickets
6. Katrina rode 21 km/hr and Naomi rode 11 km/hr
7. 17
8. 24 ways

Chapter 17, Tackling Test Tremors

Answers will vary.

Appendix, The Pascal Triangle

1. Only ones.
2. 1, 9, 36, 84, 126, 126, 84, 36, 9, 1
6. 8, 16, 32, 64, 128
7. (a) 1
   (b) $x + 1$
   (c) $x^2 + 2x + 1$
   (d) $x^3 + 3x^2 + 3x + 1$
   (e) $x^4 + 4x^3 + 6x^2 + 4x + 1$

The numbers match the rows in Pascal's Triangle.
$(x + 1)^5 = x^5 + 5x^4 + 10x^3 + 10x^2 + 5x + 1$

Appendix, Pick's Theorem

1. (a) 8   3   6
   (b) 6   0   2
   (c) 11   0 $4\frac{1}{2}$
   (d) 8   1   4
   (e) 10   3   7
   (f) 14   3   9
   (g) 6   1   3

2. (a) $15\frac{1}{2}$ square ft
   (b) 30 square ft
   (c) 25 square m
   (d) 15 square cm
   (e) 49 square inches
   (f) 16 square yd
   (g) 48 square m
   (h) 24 square m
   (i) 55 square ft
   (j) 12 square cm
   (k) 30 square in

Index